# RANDOM DIFFERENTIAL INEQUALITIES

This is Volume 150 in
MATHEMATICS IN SCIENCE AND ENGINEERING
A Series of Monographs and Textbooks
Edited by RICHARD BELLMAN, *University of Southern California*

The complete listing of books in this series is available from the Publisher
upon request.

# RANDOM DIFFERENTIAL INEQUALITIES

G. S. Ladde

and

V. Lakshmikantham

Department of Mathematics
University of Texas at Arlington
Arlington, Texas

1980

ACADEMIC PRESS

A Subsidiary of Harcourt Brace Jovanovich, Publishers

New York   London   Toronto   Sydney   San Francisco

ACADEMIC PRESS, INC.
111 Fifth Avenue, New York, New York 10003

*United Kingdom Edition published by*
ACADEMIC PRESS, INC. (LONDON) LTD.
24/28 Oval Road, London NW1   7DX


Library of Congress Cataloging in Publication Data

Ladde, G.  S.
    Random differential inequalities.

    (Mathematics in science and engineering)
    Bibliography:  p.
    Includes index.
    1.  Stochastic differential equations.
2.  Differential inequalities.  I.  Lakshmikantham,
V. , joint author.  II.  Title.  III.  Series.
QA274.23.L32     515.3'5     80–521
ISBN  0–12–432750–8


PRINTED IN THE UNITED STATES OF AMERICA

80 81 82 83     9 8 7 6 5 4 3 2 1

# CONTENTS

CHAPTER 3

## $L^p$-Calculus Approach

CHAPTER 4

## Itô–Doob Calculus Approach

## Appendix

# PREFACE

The mathematical modeling of several real-world problems leads to differential systems that involve some inherent randomness due to ignorance or uncertainties. If the randomness is eliminated, we have, of course, deterministic differential systems. Also, many important problems of the physical world are nonlinear. Consequently, the study of nonlinear random differential systems is a very important area in modern applied mathematics.

A differential system can involve random behavior in three ways: (i) random forcing functions, (ii) random initial conditions, and (iii) random coefficients. Problems in which randomness is limited to (i) and (ii) are relatively simple to investigate. However, the most interesting case of random differential equations is (iii) when combined with (i) and (ii). The objective, of course, is to discuss fundamental properties and qualitative behavior of solutions by various probabilistic modes of approach. Also, since the solutions are stochastic processes, various statistics of the solution processes are to be found.

As is well known, the theory of differential inequalities plays a crucial role in the study of deterministic differential equations. Furthermore, the theory of differential inequalities together with the concept of a Lyapunov function provides a suitable and effective mechanism for investigating a variety of qualitative aspects of solutions, including stability theory. It is natural to expect that a corresponding theory of random differential inequalities will play an equally important role in the theory of random differential equations. This is the basis for the evolution of this book.

The present book offers a systematic treatment of random differential inequalities and its theory and application—depending on the different modes of probabilistic analysis, namely, approach through sample calculus, $L^p$-mean calculus, and Itô–Doob calculus. The book is divided into four chapters. The first chapter consists of preliminary material that is required for the rest of the book. We list here needed basic concepts and results in a logical sequence. In the second chapter we develop the fundamental theory of random differential equations and inequalities in the framework of sample calculus. Chapter 3 is devoted to the treatment of random differential equations and inequalities through $p$th mean calculus. The last chapter investigates the differential equations of Itô type in the context of Itô–Doob calculus. Finally, an appendix is given to supplement the material of the book. Several examples and a carefully selected set of problems are incorporated in the body of the text.

Some of the important features of the book are the following:

(i)    inclusion of the study of random differential equations through sample calculus,

(ii)    development of the theory of random differential inequalities through various modes of probabilistic analysis and the application of these results to discuss various properties of solution processes,

(iii)    a unified treatment of stability theory through random Lypaunov functions and random comparison method, and

(iv)    stress of the role of the method of variation of parameters in the stability analysis of stochastic perturbed systems.

This monograph can be used as a textbook at the graduate level and as a reference book. A good background in probability theory and differential equations is adequate to follow the contents of this book.

We wish to express our warmest thanks to Professor Richard Bellman whose interest and enthusiastic support made this work possible. We are immensely pleased that our book appears in his series. The staff of Academic Press has been most helpful.

We thank our colleagues who participated in the seminar on stochastic differential equations at The University of Texas at Arlington. In particular, we appreciate the comments and criticism of Professors Stephen R. Bernfeld, Jerome Eisenfeld, Pat Sutherland, and Randy Vaughn. Moreover, we wish to thank Mrs. Mary Ann Crain for her excellent typing of the manuscript.

The first-mentioned author would like to acknowledge encouragement and support of his friend and colleague, Professor Clarence F. Stephens, SUNY-Potsdam. Furthermore, he would like to acknowledge the Research Foundation of the State University of New York for its continuous encouragement and support for the development of several results in this book. A large part of the book was completed while the first author was associated with SUNY-Potsdam. The preparation of this book was facilitated by U.S. Army Research Grant DAAG29-77-G0062, and we express our gratitude for the support.

# NOTATIONS AND ABBREVIATIONS

For the convenience of readers we collect below the various notations and abbreviations employed in the monograph.

Vectors (column vectors) of dimension $n$ are basically treated as $n \times 1$ matrices. All relations such as equations, inequalities, belonging to, and limits, involving random variables or functions are valid with probability one. Sometimes the symbols $x(t)$ and $x(t,\omega)$ are used interchangeably as a random function.

| | |
|---|---|
| $R^n$ | An $n$-dimensional Euclidean space with a convenient norm $\|\cdot\|$ |
| $\|\cdot\|$ | The norm of a vector or matrix |
| $R$ | The set of all deterministic real numbers or real line |
| $R_+$ | The set of all $t \in R$ such that $t \geq 0$ |
| $(\Omega, \mathscr{F}, P)$ | A probability space |
| $\mathscr{F}^n$ | The $\sigma$-algebra of Borel sets in $R^n$ |
| $\Omega \equiv (\Omega, \mathscr{F}, P)$ | A complete probability space |
| $R[\Omega, R^n]$ | The collection of all random vectors defined on a complete probability space $(\Omega, \mathscr{F}, P)$ into $R^n$ |
| $R[\Omega, R^{nm}]$ | A collection of all $n \times m$ random matrices $A = (a_{ij})$ such that $a_{ij} \in R[\Omega, R]$ |
| $\mathscr{L}^p$ | The collection of all $n$-dimensional random vectors $x$ such that $E(\|x\|^p) < \infty$ for $p \geq 1$ |
| $L^p[\Omega, R^n]$ | A collection of all equivalence classes of random vectors such that an element of an equivalence class belongs to $\mathscr{L}^p$ |
| $x^T$ | The transpose of a vector or matrix $x$ |
| a.s. | Almost surely or almost certainly |
| i.p. | In probability or stochastically |
| p.m. | $p$th mean or moment |
| $I_A$ | A characteristic or indicator function with respect to an event $A$ |

$I$      An arbitrary index set, in particular, a finite, countable set or any interval in $R$

$R[I, R[\Omega, R^n]]$      A class of random functions defined on $I$ into $R[\Omega, R^n]$

$C[E, R^n]$      The class of deterministic continuous functions defined on an open $(t,x)$ subset $E$ of $R^{n+1}$ into $R^n$

$R[[a,b], R[\Omega, R^n]]$      A collection of all $R^n$-valued separable random functions defined on $[a,b]$ with a state space $(R^n, \mathscr{F}^n)$, $a,b \in R$

$C[[a,b], R[\Omega, R^n]]$      A collection of all $R^n$-valued separable and sample continuous random functions defined on $[a,b]$ with a state space $(R^n, \mathscr{F}^n)$, $a,b \in R$

w.p. 1      With probability one

$M[[a,b], R[\Omega, R^n]]$      A collection of all random functions in $R[[a,b], R[\Omega, R^n]]$ which are product-measurable on $([a,b] \times \Omega, \mathscr{F}^1 \times \mathscr{F}, m \times P)$, where $\Omega \equiv (\Omega, \mathscr{F}, P)$ and $([a,b], \mathscr{F}^1, m)$ are a complete probability space and a Lebesgue-measurable space, respectively

$B(z,\rho)$      The set of all $x \in R^n$ such that $\|x - z\| < \rho$ for given $z \in R^n$ and positive real number $\rho$

$B(\rho)$      The set $B(z,\rho)$ with $z = 0 \in R^n$

$\bar{B}(z,\rho)$      The closure of $B(z,\rho)$

$C[R_+ \times B(z,\rho), R[\Omega, R^n]]$      A class of sample continuous $R^n$-valued random functions $f(t,x)$ whose realizations are denoted by $f(t,x,\omega)$

$M[R_+ \times B(z,\rho), R[\Omega, R^n]]$      A class of $R^n$-valued random functions such that $f(t,x(t))$ is product-measurable whenever $x(t)$ is product-measurable

$IB[I, R[\Omega, R_+]]$      The class of random functions $K \in M[I, R[\Omega, R_+]]$ such that its sample Lebesgue integral is bounded with probability one

$J$      $[t_0, t_0 + a]$, where $t_0 \in R_+$ and $a$ is a positive real number

a.e.      Almost everywhere or except a set of measure zero

det $A$      The determinant of a square matrix $A$

tr $A$      The trace of a square matrix $A$

$A^{-1}$      The inverse of a square matrix $A$

$\mu(A(\omega))$      The logarithmic norm of a random square matrix $A(\omega)$

$\sigma(A(\omega))$      The spectrum of random square matrix $A(\omega)$

$\mathscr{K}$      The class of functions $b \in C[[0,\rho), R_+]$ such that $b(0) = 0$ and $b(r)$ is strictly increasing in $r$, where $0 \leq \rho \leq \infty$

$\mathscr{V}\mathscr{K}$      The class of functions $b \in C[[0,\rho), R_+]$ such that $b(0) = 0$ and $b(r)$ is convex and strictly increasing in $r$

$\mathscr{C}\mathscr{K}$      The class of functions $a \in C[R_+ \times [0,\rho), R_+]$ such that $a(t,0) \equiv 0$ and $a(t,u)$ is concave and strictly increasing in $u$ for each $t \in R_+$

$R[[a,b],L^p[\Omega,R^n]]$   A collection of all functions defined on $[a,b]$ with values in $L^p[\Omega,\ R^n]$

$C[[a,b],L^p[\Omega,R^n]]$   A collection of functions in $R[[a,b],\ L^p[\Omega,R^n]]$ which are $L^p$-continuous on $[a,b]$

$[R_+\times L^p[\Omega,R^n],L^p[\Omega,R^n]]$   A collection of all functions $f(t,x)$ defined on $R_+\times L^p[\Omega,\ R^n]$ into $L^p[\Omega,\ R^n]$

$\mathscr{F}_t$   A sub-$\sigma$-algebra of $\mathscr{F}$ defined for $t\in R_+$

$\mathscr{Z}_t$   The smallest sub-$\sigma$-algebra of $\mathscr{F}$ generated by an $m$-dimensional normalized Wiener process $z(t)$ for $t\in R_+$

$\mathscr{Z}_t^+$   The smallest sub-$\sigma$-algebra of $\mathscr{F}$ generated by $z(t) - z(s)$ for $s\geq t\geq 0$, where $z(t)$ is an $m$-dimensional normalized Wiener process

$V_x\equiv\dfrac{\partial V}{\partial x}\ (t,x)$   An $N\times n$ Jacobian matrix of $V(t,x)$, where $V\in C[R_+\times R^n,\ R^N]$

$V_{xx}\equiv\dfrac{\partial^2 V}{\partial x^2}\ (t,x)$   An $n\times n$ Hessian matrix of $V\in C[R_+\times R^n,\ R^N]$ whose elements $(\partial^2 V/\partial x_i\partial x_j)\ (t,x)$ are $N$-dimensional vectors

$M_2[a,b]$   A set of all nonanticipating $n\times m$ matrix or $n$-vector functions $G$ defined on $[a,b]$ into $R[\Omega R^{nm}]$ or $R[\Omega,R^n]$ such that the sample Lebesgue integral $\int_a^b\|G(s,\omega)\|^2\ ds$ exists with probability one

m.s.   Mean square

# CHAPTER 1

# PRELIMINARY ANALYSIS

## 1.0. INTRODUCTION

This chapter is essentially introductory in nature. Its main purpose is to introduce some basic probabilistic concepts that are essential to the study of random differential equations. We list some known results from standard textbooks and sketch some useful results that are not so well known. Sections 1.1 and 1.2 are concerned with events, probability measure, random variables, expectations, and distribution functions. Section 1.3 is devoted to convergence of random sequences depending on various modes of convergence. Several important and well-known theorems about various kinds of convergence are stated. Furthermore, some results concerning the weak convergence of measures are also included. A brief discussion about conditional probabilities and expectations is given in Section 1.4. Section 1.5 surveys certain fundamental notions and results in the general theory of stochastic processes. We sketch an important concept due to Doob, namely, the concept of separability of random processes in Section 1.6. Finally, we collect in Section 1.7 some basic deterministic comparison results.

## 1.1. EVENTS AND PROBABILITY MEASURE

Let us consider a random experiment $E$ whose outcomes, the elementary events $\omega$, are elements of a set $\Omega$. $\Omega$ is called a sample space. Let $\mathscr{F}$ denote a $\sigma$-algebra of subsets of the sample space $\Omega$. Elements of $\mathscr{F}$ are called events. A real-valued set function $P$ defined on $\mathscr{F}$ is called a probability measure or simply a probability if

(i)  $P(A) \geq 0, \forall A \in \mathscr{F}$ (nonnegativity);
(ii)  $P(\Omega) = 1$ (normed finiteness);
(iii)  $P(\bigcup_{n=1}^{\infty} A_n) = \sum_{n=1}^{\infty} P(A_n)$ for $A_n \in \mathscr{F}$ and $n \geq 1$, $A_n \cap A_m = \varnothing$
$(n \neq m)$ (semiadditivity).

The triplet $(\Omega, \mathscr{F}, P)$ is called a probability space. A subset of an event of zero probability is called a null event or null set. A probability space $(\Omega, \mathscr{F}, P)$ is said to be complete if every null event is an event. If $(\Omega, \mathscr{F}, P)$ is not complete, $P$ can be uniquely extended to the $\sigma$-algebra $\mathscr{F}$ generated by $\mathscr{F}$ and its null events. This procedure is called completion.

Let $\mathscr{F}^n$ denote the $\sigma$-algebra of Borel sets in $R^n$ generated by the $n$-dimensional intervals

$$I^n = \{x \in R^n : \infty < x_i < q_i, q_i \in R, \forall i = 1, 2, \ldots, n\}.$$

Let $(\Omega_i, \mathscr{F}_i, P_i)$ be probability spaces for $i = 1, 2, \ldots, n$. We construct a probability space $(\Omega, \mathscr{F}, P)$ in the following way: We let $\Omega$ be the Cartesian product

$$\Omega = \Omega_1 \times \Omega_2 \times \cdots \times \Omega_n,$$

consisting of all $n$-tuples $\omega = (\omega_1, \omega_2, \ldots, \omega_n)$ such that $\omega_i \in \Omega_i$. In $\Omega$, we define the product $\sigma$-algebra of subsets of $\Omega$

$$\mathscr{F} = \mathscr{F}_1 \times \mathscr{F}_2 \times \cdots \times \mathscr{F}_n$$

as the $\sigma$-algebra generated by the cylinder sets in

$$A = A_1 \times A_2 \times \cdots \times A_n, \qquad A_i \in \mathscr{F}_i.$$

Note that $\mathscr{F}$ is the smallest $\sigma$-algebra with respect to which the projections $p_i : \Omega \to \Omega_i$ defined by $p_i(\omega) = \omega_i$ are measurable. The probability on $\Omega$ is defined by

$$P = P_1 \times P_2 \times \cdots \times P_n,$$

with the property that

$$P(A_1 \times A_2 \times \cdots \times A_n) = P(A_1)P(A_2)\cdots P(A_n) \qquad \text{for all} \quad A_i \in \mathscr{F}_i.$$

This probability $P$ is called the product probability on $\Omega$. Hence the triplet $(\Omega, \mathscr{F}, P)$ is called the product probability space.

Let $(\Omega, \mathscr{F}, P)$ be a probability space. The events $A_1, A_2, \ldots, A_n$ are said to be independent (statistically) if

$$P\left(\bigcap_{i=1}^{n} A_i\right) = P(A_1)P(A_2)\cdots P(A_n).$$

Sub-$\sigma$-algebras $\mathscr{F}_1, \mathscr{F}_2, \ldots, \mathscr{F}_n$ of $\mathscr{F}$ are said to be independent if the above relation holds for every choice of events $A_i \in \mathscr{F}_i$ for $i = 1, 2, \ldots, n$.

## 1.2. RANDOM VARIABLES, DISTRIBUTION FUNCTIONS, AND EXPECTATIONS

Let $\Omega \equiv (\Omega, \mathscr{F}, P)$ be a complete probability space. A function $x : \Omega \to R^n$ is said to be a $R^n$-valued measurable function if the preimages of measurable sets in $R^n$ are measurable sets in $\Omega$, i.e., if $I^n \in \mathscr{F}^n$,

$$x^{-1}(I^n) = \{\omega : \infty < x_i(\omega) < q_i, \forall i = 1, 2, \ldots, n\} \in \mathscr{F}.$$

This measurable function $x$ is called a random variable or random vector. The collection of all random vectors is denoted by $R[\Omega, R^n]$. Similarly, an $n \times m$ matrix function $A : \Omega \to R^{nm}$ defined by $A(\omega) = (a_{ij}(\omega))$ is called a random matrix if the elements $a_{ij}(\omega)$ are real-valued random variables.

Let $x$ be a random vector. The function defined by

$$\begin{aligned} F(a) &= F(a_1, a_2, \ldots, a_n) \\ &= P\{\omega : x_1(\omega) \le a_1, \ldots, x_n(\omega) \le a_n\} = P[x \le a] \end{aligned} \quad (1.2.1)$$

is called the distribution function. It is also called the joint distribution function of the $n$ real-valued random variables $x_1, x_2, \ldots, x_n$, which are components of $x$. The function $F(a)$ defined by (1.2.1) is monotone non-decreasing and continuous from the right at $a$, with

$$\lim_{a_i \to -\infty} F(a_1, a_2, \ldots, a_n) = 0, \qquad i = 1, 2, \ldots, n,$$

and

$$\lim_{a_1, \ldots, a_n \to \infty} F(a_1, a_2, \ldots, a_n) = 1.$$

The function obtained from the distribution function $F(a)$ of $n$ real-valued random variables by replacing in $F(a_1, a_2, \ldots, a_n)$ some of its arguments with $\infty$, say $a_1, a_2, \ldots, a_k$, is called the marginal distribution of the random variables $x_{k+1}, x_{k+2}, \ldots, x_n$. Note that the distribution function $F$ defines a probability measure

$$P_x(A) = \int_A d_a F(a) \quad (1.2.2)$$

(Lebesgue–Stieltjes measure in $R^n$). A function $f : R^n \to R_+$ is said to be the joint density or simply density function of a random vector $x$ whose distribution function is defined by

$$F(a_1, a_2, \ldots, a_n) = \int_{-\infty}^{a_1} \cdots \int_{-\infty}^{a_n} f(u_1, u_2, \ldots, u_n) \, du_1 \cdots du_n.$$

Let $x$ be a $n$-dimensional random vector. A random variable $x$ is said to be $P$-integrable if the integral $\int_\Omega x \, dP$ is finite, and it is called the expectation or mean of $x$, and we write

$$E(x) = \int_\Omega x P(d\omega) = \begin{bmatrix} \int_\Omega x_1 P(d\omega) \\ \vdots \\ \int_\Omega x_n P(d\omega) \end{bmatrix}.$$

Similarly, the expectation of a random matrix $A$ is defined by

$$\int_\Omega A P(d\omega) = E(A) = [E(a_{ij})].$$

The expectation of a random variable is also defined in terms of its distribution function, i.e.,

$$E(x) = \int_{R^n} a \, d_a F(a),$$

where $F$ is the distribution function of $x$ and the above integral is the integral of a function $g(a) = a$ over the probability space $(R^n, \mathscr{F}^n, P_x)$ and is known as the Lebesgue–Stieltjes integral. Note that $E(x)$ satisfies the following properties:

   (i)   $E(x + y) = E(x) + E(y)$ (additivity);
   (ii)  $E(cx) = cE(x)$ (homogeneity) for any real constant $c$;
   (iii) $x \geq y$ implies $E(x) \geq E(y)$ (order preservation).

Suppose that for $p \geq 1$, $\mathscr{L}^p = \mathscr{L}^p(\Omega, \mathscr{F}, P) = \{x : x \text{ is an } n\text{-dimensional}$ random variable and $E(\|x\|^p) < \infty\}$. Then $\mathscr{L}^p$ is a linear space. Also, if $L^p[(\Omega, \mathscr{F}, P), R^n] = \{\bar{x} : \bar{x} \text{ is an equivalence class of random variable } x \text{ almost}$ surely (a.s.) and $x \in \mathscr{L}^p\}$, then $L^p[\Omega, R^n] \equiv L^p[(\Omega, \mathscr{F}, P), R^n]$ is a Banach space with respect to the norm

$$\|\bar{x}\|_p = [E(\|\bar{x}\|^p)]^{1/p}.$$

In particular, for $p = 2$, it is a Hilbert space with scalar product

$$\langle x, y \rangle = E(x^T y) = \sum_{i=1}^{n} E(x_i y_i),$$

where $x$ and $y$ are equivalent classes of random variables and superscript T stands for transpose of a matrix or vector. Further note that when $p = \infty$,

$$\|x\|_\infty = \sup\{r : P(\|x\| > r) > 0\}.$$

One can easily show that $\|x\|_p$ verifies the following properties:

   (i)   $0 \leq \|x\|_1 \leq \|x\|_r \leq \|x\|_p \leq \|x\|_\infty \leq \infty$ $(1 \leq r \leq p \leq \infty)$;
   (ii)  $P(\|x\| \neq 0) = 0$ if and only if $\|x\|_p = 0$ $(1 \leq p \leq \infty)$;

(iii)  $|\langle x, y \rangle| \leq \|x\|_p \|y\|_q$  for  $p > 1$ ,  $(1/p) + (1/q) = 1$ , for  $x \in L^p$ ,  $y \in L^q$
(Hölder's inequality);
(iv)  $\|x + y\|_p \leq \|x\|_p + \|y\|_p$ ,  $p \geq 1$ ,  $x, y \in L^p$  (Minkowski's inequality).

**Theorem 1.2.1**[1]  (Jensen's inequality).  If  $\phi$  is a real-valued continuous
and concave function defined on a convex domain  $D \subseteq R^n$ , then

$$E[\phi(x)] \leq \phi(E(x)).$$

For  $p > 0$ ,  $\varepsilon > 0$ ,  $x \in L^p$ , we have

$$P(\|x\| \geq \varepsilon) < \varepsilon^{-p} \|x\|_p^p \quad \text{(Markov's or Chebyshev's inequality)}.$$

For  $n = 1$ , the number

$$\operatorname{var}(x) = E((x - E(x))^2) = \sigma^2(x), \qquad x \in L^2,$$

is called the variance of  $x$ , the number  $\sigma = [\operatorname{var}(x)]^{1/2}$  is called the standard
deviation of  $x$ , the number  $E(x^k)$  is called the  $k$ th moment of  $x$ , the number
 $E((x - E(x))^k)$  is called the  $k$ th central moment of  $x$ , and the number

$$\operatorname{cov}(x, y) = E((x - E(x))(y - E(y)))$$

is called the covariance of  $x$  and  $y$  for  $x, y \in L^2$ . For  $n$ -dimensional random
vectors  $x, y$ , the symmetric nonnegative definite  $n \times n$  matrix

$$\operatorname{cov}(x, y) = E((x - E(x))(y - E(y))^{\mathrm{T}}) = (\operatorname{cov}(x_i, y_j))$$

is called the covariance matrix of random vectors  $x$  and  $y$ . Let  $x$  be an  $n$ -
dimensional random vector. The characteristic function of a random vector  $x$
is defined by

$$Q(a) = Q_x(a) = E(e^{ia^{\mathrm{T}}x}) = \int_{R^n} e^{ia^{\mathrm{T}}x} d_x F(x)$$

for  $a \in R^n$ . If  $F(a)$  is absolutely continuous with density  $f(a)$ , then the above
equation reduces to

$$Q(a) = \int_{R^n} e^{ia^{\mathrm{T}}x} f(x) \, dx,$$

and  $f(a)$  can be obtained from  $Q(a)$  by the inversion formula of the Fourier
integral,

$$f(x) = \frac{1}{2\pi} \int_{R^n} e^{-ia^{\mathrm{T}}x} Q(a) \, da.$$

[1] Neveu [74]

## 1.3. CONVERGENCE OF RANDOM SEQUENCES

Let $x, x_1, x_2, \ldots$ be $n$-dimensional random vectors defined on a probability space $(\Omega, \mathscr{F}, P)$.

**Definition 1.3.1.** A sequence of random variables $\{x_n\}$ is said to converge *almost surely* or *almost certainly* to $x$ if there exists an event $N \in \mathscr{F}$ such that $P(N) = 0$, and for every $\omega \in \Omega \backslash N$

$$\lim_{n \to \infty} \|x_n(\omega) - x(\omega)\| = 0,$$

and we write

$$x_n \xrightarrow[n \to \infty]{\text{a.s.}} x.$$

**Definition 1.3.2.** A sequence $\{x_n\}$ is said to converge *in probability* or *stochastically* to $x$ if for every $\varepsilon > 0$,

$$\lim_{n \to \infty} P\{\omega : \|x_n(\omega) - x(\omega)\| > \varepsilon\} = 0,$$

and we denote it

$$x_n \xrightarrow[n \to \infty]{\text{i.p.}} x.$$

**Definition 1.3.3.** A sequence of random variables $\{x_n\}$ is said to converge in the *pth mean or moment* $(p > 0)$ to $x$ if

$$\lim_{n \to \infty} E[\|x_n - x\|^p] = 0,$$

and we use the notation

$$x_n \xrightarrow[n \to \infty]{\text{p.m.}} x.$$

**Remark 1.3.1.** In Definition 1.3.3, for $p = 1$, the $p$th convergence is simply convergence in the mean, and for $p = 2$, the $p$th convergence is simply convergence in the mean square or in the quadratic mean.

Let $\{F_n\}$ and $F$ denote the distribution functions of $\{x_n\}$ and $x$, respectively.

**Definition 1.3.4.** A sequence of random variables $\{x_n\}$ is said to converge *in distribution* to $x$ if

$$\lim_{n \to \infty} F_n(x) = F(x)$$

at every point at which $F$ is continuous or

$$\lim_{n \to \infty} Q_n(u) = Q(u) \qquad \text{for all} \quad u \in R^n,$$

where $Q_n$ and $Q$ are characteristic functions of $x_n$ and $x$, respectively.

**Remark 1.3.2.** These convergence concepts are related with each other as follows:

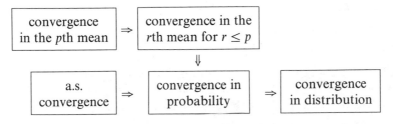

**Remark 1.3.3.** (1) If there exists a number $A$ such that $\|x_n\| \leq A$ a.s. for $n \geq 1$ ($x_n$ is a.s. uniformly bounded), then $\{x_n\}$ converges in $p$th mean.

(2) If $x_n \leq x_{n+1}$ a.s. for all $n \geq 1$ or if $x_n \geq x_{n+1}$ a.s. for all $n$, then each one of the three convergences (in probability, a.s., and quadratic mean) implies the other two.

(3) A sequence converges in probability if and only if every subsequence of it contains an a.s. convergent subsequence.

We present some well-known results that are useful in our study.

**Theorem 1.3.1** (Monotone convergence). Let $\{x_n\}$ be an increasing sequence of nonnegative random variables converging a.s. to $x$. Then $E(x) = \lim_{n \to \infty} E(x_n)$.

**Theorem 1.3.2** (Fatou–Lebesgue lemma). Let $x_n$ be a sequence of random variables. Suppose that there exist integrable random variables $x$, $y$ such that $x_n(\omega) \geq x(\omega)$ and $x_n(\omega) \leq y(\omega)$ a.s. for all $n$. Then

$$\liminf_{n \to \infty} E(x_n) \geq E\left( \liminf_{n \to \infty} x_n \right),$$

and

$$\limsup_{n \to \infty} E(x_n) \leq E\left( \limsup_{n \to \infty} x_n \right).$$

**Theorem 1.3.3** (Dominated convergence). Let $\{x_n\}$ be a sequence of random variables converging almost surely to $x$. Suppose that there exists an integrable random variable $y$ such that $\|x_n(\omega)\| \leq \|y(\omega)\|$ a.s. for all $n$. Then

$$\lim_{n \to \infty} E(x_n) = E(x).$$

**Theorem 1.3.4** (Borel–Cantelli lemma). For any arbitrary sequence of events $\{A_n\}$, $\sum_{n=1}^{\infty} P(A_n) < \infty$ implies $P(\limsup_{n \to \infty} A_n) = 0$. Moreover, if the sequence $\{A_n\}$ is independent, $\sum_{n=1}^{\infty} P(A_n) = \infty$ implies $P(\limsup_{n \to \infty} A_n) = 1$.

Let $\{x_n\}$ be a sequence of real-valued mutually independent random variables defined on $(\Omega, \mathscr{F}, P)$. Without loss of generality, we assume that $E(x_n) = 0$, $\sigma_n^2 = E(x_n^2)$ for all $n \geq 1$. Let us set

$$s_n = \sum_{i=1}^{n} x_i.$$

The law of convergence of $s_n$ is said to be the *weak law of convergence* if the convergence is in probability; the law of convergence of $s_n$ is said to be the *strong law of convergence* if the convergence is a.s.

**Theorem 1.3.5** (Weak law of large numbers). Let $\{x_n\}$ be a sequence of real-valued mutually independent random variables with $E(x_i) = 0$ and finite variances $\sigma_i^2$. Then

$$(1/n)\, s_n \to 0 \qquad \text{in the mean square as} \quad n \to \infty$$

iff

$$\sum_{n=1}^{\infty} \sigma_n^2 < \infty.$$

**Theorem 1.3.6** (Strong law of large numbers). Let $\{x_n\}$ be a sequence of real-valued mutually independent random variables with $E(x_i) = 0$, $\sigma_i^2 < \infty$. Then

$$\sum_{i=1}^{\infty} \sigma_i^2 / i^2 < \infty$$

implies $(1/n)\, s_n \to 0$ a.s. as $n \to \infty$.

**Theorem 1.3.7** (Central limit theorem). Let $\{x_n\}$ be a sequence of real-valued independent random variables with a common distribution function, having finite mean $\mu$ and variance $\sigma^2$. Then $(s_n - n\mu)/\sigma\sqrt{n}$ is asymptotically Gaussian with expectation 0 and variance 1,

$$\lim_{n \to \infty} P\left[\omega : \frac{s_n - n\mu}{\sqrt{n}} \leq \sigma\lambda\right] = \frac{1}{\sqrt{2\pi}} \int_{-\infty}^{\lambda} e^{-s^2/2}\, ds$$

uniformly in $\lambda$.

Let us recall that $\mathscr{F}^n$ is a $\sigma$-algebra of Borel sets in the $n$-dimensional Euclidean space $R^n$. Let $P_1$ and $P_2$ be probability measures defined on $(R^n, \mathscr{F}^n)$. For every closed subset $F$ of $R^n$, let $\varepsilon_{12}$ be the infimum of $\varepsilon > 0$ such that

$$P_1(F) < P_2(O_\varepsilon(F)) + \varepsilon, \tag{1.3.1}$$

where $O_\varepsilon(F)$ is the $\varepsilon$-neighborhood of $F$. By interchanging the roles of $P_1$ and $P_2$ in (1.3.1), $\varepsilon_{21}$ can be defined analogously.

**Definition 1.3.5.** The Prohorov distance $D(P_1, P_2)$ is defined by

$$D(P_1, P_2) = \max(\varepsilon_{12}, \varepsilon_{21}).$$

We note that the set of probability measures together with distance $D$ in Definition 1.3.5 is a complete separable metric space and $D$-convergence is equivalent to weak convergence of measures. For details, see [75].

Let $x, y \in R[\Omega, R^n]$, and let $P_x, P_y$ be corresponding probability measures defined on $(R^n, \mathscr{F}^n)$. The Prohorov distance between their probability laws, that is, $P_x, P_y$, is denoted by $D(x, y) = D(P_x, P_y)$. We notice that $D(x, y) = 0$ means that $x$ and $y$ have the same probability measure or law.

We remark that $P\{\omega: \lim_{n \to \omega} \|x_n(\omega) - x(\omega)\| = 0\} = 1$ implies $x_n$ is a $D$-Cauchy sequence. The converse is also true in the following sense.

**Theorem 1.3.8** (Skorokhod's theorem). Let $x_n \in R[\Omega, R^n]$ be a $D$-Cauchy sequence of random variables. Then one can construct a sequence of random variables $y_n \in R[\Omega, R^n]$ and a random variable $y \in R[\Omega, R^n]$ such that

$$D(y_n, x_n) = 0 \qquad \text{and} \qquad P\{\omega: \|y_n(\omega) - y(\omega)\| = 0\} = 1.$$

**Definition 1.3.6.** A collection $S = \{x_\alpha: \alpha \in \Lambda\} \subset R[\Omega, R^n]$ is said to be totally $D$-bounded if every infinite sequence $\{x_{n_\alpha}\} \subseteq S$ has a $D$-Cauchy subsequence, where $\Lambda$ is an index set.

The following theorem gives a necessary and sufficient condition for a set in $R[\Omega, R^n]$ to be totally $D$-bounded.

**Theorem 1.3.9** (Prohorov's theorem). Let $S \subset R[\Omega, R^n]$. Then $S$ is totally $D$-bounded in $R[\Omega, R^n]$ iff for every $\varepsilon > 0$, there exists a compact subset $K_\varepsilon$ of $R^n$ such that

$$P(x \in K_\varepsilon) > 1 - \varepsilon$$

for every $x \in S$.

We remark that Theorems 1.3.8 and 1.3.9 are valid for a collection of random variables that are defined on a complete probability space with values in a complete separable metric space $(M, d)$ with a distance $d$. In the light of this, we present a result which shows that the direct product of a finite number of totally $D$-bounded sets in a complete separable metric space is totally $D$-bounded in a direct product metric space. Let $(M_i, d_i), i = 1, 2, \ldots,$ $n$, be complete separable metric spaces. Then the direct product $(M, d) =$ $(M_1, d_1) \times (M_2, d_2) \times \cdots \times (M_n, d_n)$ is also a complete separable metric space with $d = \sum_{i=1}^{n} d_i$, $M = M_1 \times M_2 \times \cdots \times M_n$. Let $S = \{x_\alpha = (x_{\alpha 1}, x_{\alpha 2}, \ldots, x_{\alpha n}) \in R[\Omega, (M, d)]: \alpha \in \Lambda\}$ be a subset of $R[\Omega, (M, d)]$. Using Prohorov's theorem and recalling the fact that the direct product of compact sets is compact and that the projection of a compact set is compact, we can immediately see the validity of the following result.

**Lemma 1.3.1.**   The set $S \subset R[\Omega, (M, d)]$ is totally $D$-bounded if and only if $S_i = \{x_{\alpha i} \in R[\Omega, (M_i, d_i)] : \alpha \in \Lambda\}$ is totally $D$-bounded for every $i = 1, 2, \ldots, n.$

## 1.4.  CONDITIONAL PROBABILITIES AND EXPECTATIONS

Let $(\Omega, \mathcal{F}, P)$ denote a probability space and let $\mathcal{F}_1 \subset \mathcal{F}$ denote a sub-$\sigma$-algebra of $\mathcal{F}$. Let $x \in L^1[(\Omega, \mathcal{F}, P), R^n]$. The probability space $(\Omega, \mathcal{F}_1, P)$ is a coarsening of $(\Omega, \mathcal{F}, P)$, and $x$ is, in general, no longer $\mathcal{F}_1$-measurable.

**Definition 1.4.1.**   The conditional expectation of $x$ relative to $\mathcal{F}_1$ is an $\mathcal{F}_1$-measurable random variable, denoted by $E(x|\mathcal{F}_1)$, and is defined by

$$\int_A E(x|\mathcal{F}_1)P(d\omega) = \int_A xP(d\omega), \qquad A \in \mathcal{F}_1.$$

According to Radon–Nikodym theorem, $E(x|\mathcal{F}_1)$ exists and is almost-surely unique.

**Theorem 1.4.1**  (Radon–Nikodym).   Let $(\Omega, \mathcal{F}, P)$ be a probability space and let $u$ be an absolutely continuous measure on $(\Omega, \mathcal{F})$. Then there exists a finite-valued measurable function $x$ on $\Omega$ such that

$$u(A) = \int_A xP(d\omega), \qquad A \in \mathcal{F}.$$

Furthermore, $x$ is unique up to sets of $P$-measure zero.

Conditional expectations possess the following important properties with probability one:

(i)   For $\mathcal{F} = \{\phi, \Omega\}$, $E(x|\mathcal{F}_1) = E(x)$.
(ii)   If $x \geq 0$, then $E(x|\mathcal{F}_1) \geq 0$.
(iii)   If $x$ is $\mathcal{F}_1$-measurable, then $E(x|\mathcal{F}_1) = x$.
(iv)   If $x = a$, then $E(x|\mathcal{F}_1) = a$.
(v)   For $\mathcal{F}_1 \supset \mathcal{F}$, if $E[x]$ exists, then $E[E[x|\mathcal{F}_1]] = E[x]$.
(vi)   If $c_1, c_2, \ldots, c_n$ are constants, then

$$E\left(\sum_{i=1}^n c_i x_i \Big| \mathcal{F}_1\right) = \sum_{i=1}^n c_i E(x_i|\mathcal{F}_1).$$

(vii)   If $x \leq y$, then

$$E(x|\mathcal{F}_1) \leq E(y|\mathcal{F}_1) \qquad \text{and} \qquad |E(x|\mathcal{F}_1)| \leq E(|x||\mathcal{F}_1).$$

(viii)   If $x, \mathscr{F}_1$ are independent, then

$$E(x|\mathscr{F}_1) = E(x),$$

(ix)   For $\mathscr{F}_1 \subset \mathscr{F}_2 \subset \mathscr{F}$, then

$$E(E(x|\mathscr{F}_2)|\mathscr{F}_1) = E(E(x|\mathscr{F}_1)|\mathscr{F}_2) = E(x|\mathscr{F}_1).$$

Furthermore, conditional expectation also has the convergence property of expectation. The results corresponding to Theorems 1.3.1–1.3.3 are also valid.

Let $(\Omega, \mathscr{F}, P)$ be a probability space and $A \in \mathscr{F}, \mathscr{F}_1 \subset \mathscr{F}$.

**Definition 1.4.2.**   The conditional probability $P(A|\mathscr{F}_1)$ of an event $A$ under the condition $\mathscr{F}_1 \subset \mathscr{F}$ is defined by

$$P(A|\mathscr{F}_1) = E(I_A|\mathscr{F}_1),$$

where $I_A$ is defined as

$$I_A(\omega) = \begin{cases} 1 & \text{if} \quad \omega \in A, \\ 0 & \text{if} \quad \omega \notin A. \end{cases}$$

Let $x$ be an $n$-dimensional random variable on $(\Omega, \mathscr{F}, P)$. Consider the conditional probability

$$P(x \in B|\mathscr{F}_1) = P(\{\omega : x(\omega) \in B\}|\mathscr{F}_1),$$

where $B \in \mathscr{F}^n$. There exists a function $p(\omega, B)$ defined on $\Omega \times \mathscr{F}^n$ with the following properties: For fixed $\omega \in \Omega$, the function $p(\omega, \cdot)$ is a probability on $\mathscr{F}^n$; for fixed $B$, the function $p(\cdot, B)$ is a version of $P(x \in B|\mathscr{F}_1)$, that is, $p(\cdot, B)$ is $\mathscr{F}_1$-measurable and

$$P(A \cap \{\omega : x(\omega) \in B\}) = \int_A p(\omega, B) \, dP(\omega), \qquad A \in \mathscr{F}_1.$$

Such a function $p$ is called the conditional probability distribution of $x$ for given $\mathscr{F}_1$. For $f(x) \in L^1$,

$$E(f(x)|\mathscr{F}_1) = \int_{R^n} f(x) p(\omega, dx).$$

If $\mathscr{F}_1$ is the $\sigma$-algebra generated by the random variable $y$, then

$$E(x|\mathscr{F}_1) = E(x|y),$$

and

$$E(f(x)|y = a) = \int_{R^n} f(x) p(a, dx),$$

where

$$P(x \in B|y = a) = p(a, B).$$

If $p(a, \cdot)$ has density $h(x, a)$ in $R^n$, this density is called the conditional density of $x$ under the condition $y = a$:

$$E(f(x)|y = a) = \int_{R^n} f(x)h(x, a)\, dx \Big/ \int_{R^n} h(x, a)\, dx.$$

## 1.5.   RANDOM PROCESSES

In this section, we shall present a very brief survey of fundamental notions and results in the general theory of stochastic processes.

Let $I$ denote an arbitrary index set, and let $\Omega \equiv (\Omega, \mathcal{F}, P)$ denote a probability space and $(R^n, \mathcal{F}^n)$ denote the state space.

**Definition 1.5.1.**   A family $\{x(t), t \in I\}$ of $R^n$-valued random variables defined on a probability space $(\Omega, \mathcal{F}, P)$ is called a stochastic process or random process or random function with parameter set $I$ and state space $(R^n, \mathcal{F}^n)$. The class of random functions defined on $I$ into $R[\Omega, R^n]$ is denoted by $R[I, R[\Omega, R^n]]$.

If $I$ is a finite or countable set, then we are dealing with finitely many random variables or a random sequence. In what follows, $I$ is always one of the intervals of the type $[t_0, t_1)$, $[t_0, t_1]$, $(t_0, t_1)$, and $(t_0, t_1]$, where $-\infty \leq t_0 \leq t_1 \leq \infty$.

If $\{x(t) : t \in I\}$ is a stochastic process, then for each $t \in I$, $x(t, \cdot)$ is an $R^n$-valued random variable which is called a section at $t$ of $x(t)$, whereas for each $\omega \in \Omega$, $x(\cdot, \omega)$ is an $R^n$-valued function defined on $I$. Hence $x(\cdot, \omega)$ is an element of the product space $(R^n)^I$. It is called a section at $\omega$ or realization or trajectory or path of the stochastic process.

The finite-dimensional distribution of a stochastic process $\{x(t) : t \in I\}$ is given by

$$P\{x(t) \leq x\} = F_t(x),$$

that is, $P\{x(t_1) \leq x_1, \ldots, x(t_n) \leq x_n\} = F_{t_1, t_2, \ldots, t_n}(x_1, x_2, \ldots, x_n)$. where $t$, $t_i \in I, x, x_i \in R^n$ for $i = 1, 2, \ldots, n$. Note that this distribution function satisfies the following two conditions:

(i)   Condition of symmetry.   If $i_1, i_2, \ldots, i_n$ is a permutation of numbers $\{1, 2, \ldots, n\}$, then for arbitrary instances and $n \geq 1$,

$$F_{t_{i_1}, t_{i_2}, \ldots, t_{i_n}}(x_{i_1}, x_{i_2}, \ldots, x_{i_n}) = F_{t_1, t_2, \ldots, t_n}(x_1, x_2, \ldots, x_n).$$

(ii)   Condition of compatibility. For $m < n$ and arbitrary $t_{m+1}, \ldots, t_n \in I$,

$$F_{t_1, t_2, \ldots, t_m, t_{m+1}, \ldots, t_n}(x_1, x_2, \ldots, x_m, \infty, \ldots, \infty)$$
$$= F_{t_1, t_2, \ldots, t_m}(x_1, x_2, \ldots, x_m).$$

In practice, one knows a family of distributions $P_{t_1,\ldots,t_n}(B_1, B_2, \ldots, B_n)$ or their distribution functions $F_{t_1, t_2, \ldots, t_n}(x_1, x_2, \ldots, x_n)$ which satisfies the symmetry and compatibility conditions.

**Theorem 1.5.1** (Kolmogorov's fundamental theorem).  For every family of distribution functions that satisfies the symmetry and compatibility conditions, there exists a probability space $(\Omega, \mathscr{F}, P)$ and a stochastic process $\{x(t): t \in I\}$ having the given finite-dimensional distributions.

In fact, we take $\Omega = (R^n)^I$,

$$x(t, \omega) = \text{value of}\quad \omega \quad \text{at}\quad t. \tag{1.5.1}$$

Note that $\{x(t, \omega), \omega \in (R^n)^I, t \in I\}$ as defined in (1.5.1) is called the coordinate function, $x(t, \omega)$ is the $t$-th coordinate of $\omega$. Sets of the form

$$\{\omega : x(t_1, \omega) \in B_1, x(t_2, \omega) \in B_2, \ldots, x(t_n, \omega) \in B_n\},$$

where $B_1, B_2, \ldots, B_n$ are $n$-dimensional Borel sets, are called cylinder sets. Let $\mathscr{F}_x$ be the minimal $\sigma$-algebra containing the cylinder sets. The probability determined by the above cylinder sets can be uniquely extended on $\mathscr{F}_x$.

**Definition 1.5.2.**  Two processes with the same state space and index set are said to be equivalent iff their finite dimensional distributions are identical.

In the following, we present a few well-known examples of random functions.

**Definition 1.5.3.**  A stochastic process $x \in R[I, R[\Omega, R^n]]$ is said to be a process with independent increments if for all $t_0 < t_1 < \cdots < t_k$ in $I$, the quantities $x(t_0), x(t_1) - x(t_2), \ldots, x(t_k) - x(t_{k-1})$ are mutually independent.

**Definition 1.5.4.**  A process $x(t)$ with independent increments with values in $R$ is said to be a Poisson process $P(\lambda(t))$ if the increments $x(t) - x(s)$ are distributed according to the Poisson law with parameter $\lambda(t) - \lambda(s)$, where $\lambda(t)$ is a nondecreasing real-valued function defined on $I$. In the case $\lambda(t) = \lambda t$, $\lambda > 0$, the process $x(t)$ is said to be a homogeneous Poisson process $P(\lambda)$.

Thus for the process $P(\lambda(t))$, we have

$$P\{x(t) - x(s) = k\} = \frac{1}{k!}[\lambda(t) - \lambda(s)]^k \exp[\lambda(s) - \lambda(t)], \qquad k \in N,$$

where $N$ is a set of natural numbers,

$$E\{x(t)\} = \lambda(t) = \text{var}(x(t)),$$

and

$$Q_t(\theta) = \exp\{(\lambda(t) - \lambda(s))(e^{i\theta} - 1)\}.$$

**Definition 1.5.5.**  A process $x \in R[I, R[\Omega, R^n]]$ is said to be Gaussian if it has Gaussian (normal) distribution

$$P\{x(t) < a\} = \frac{1}{\sqrt{(2\pi)^n \det(V(t))}} \int_{-\infty}^{a_1} \cdots \int_{-\infty}^{a_n} \exp[-\tfrac{1}{2}(u - m(t))^{\mathrm{T}}$$

$$\times (V(t))^{-1}(u - m(t))] \, du_1 \cdots du_n,$$

that is, $x(t)$ has a density function $f$ such that

$$f(a) = \frac{1}{\sqrt{(2\pi)^n \det(V(t))}} \exp[-\tfrac{1}{2}(a - m(t))^{\mathrm{T}}(V(t))^{-1}(a - m(t))],$$

where $m(t)$ and $V(t)$ are the mean $n$-vector and variance $n \times n$ symmetric matrix functions on $I$, respectively. In the case $m(t) = mt$ and $V(t) = Dt$, where $D$ is a symmetric nonnegative definite matrix and $m$ is a constant $n$-vector, the process is called a homogenous Gaussian process.

Thus for the Gaussian process, the characteristic function is given by

$$Q_t(a) = \exp[ia^{\mathrm{T}} m(t) - \tfrac{1}{2} a^{\mathrm{T}} V(t) a],$$

where $a \in R^n$, $t \in I$.

**Definition 1.5.6.**  A Gaussian process with independent increments is called a Wiener process. In particular, a Gaussian process with independent increments is called a process of Brownian motion or normalized Wiener process if

$$E[x(t) - x(s)] = 0,$$

$$E[(x(t) - x(s))^{\mathrm{T}}(x(t) - x(s))] = I|t - s| \qquad \text{for} \quad t, s \in I,$$

where $I$ is an $n \times n$ identity matrix.

Thus for the Wiener process, we have

$$P(x(t) < x) = \frac{1}{\sqrt{(2\pi)^n t}} \int_{-\infty}^{x_1} \cdots \int_{-\infty}^{x_n} \exp\left[-\frac{1}{2t} u^{\mathrm{T}} u\right] du_1 \, du_2 \cdots du_n$$

and the characteristic function

$$Q_t(u) = \exp\left[-\frac{1}{2t} u^{\mathrm{T}} u\right].$$

**Definition 1.5.7.**  A random process $x \in R[I, R[\Omega, R^n]]$ is said to be a Markov process if for any increasing collection $t_1 < t_2 < \cdots < t_n$ in $I$ and $B \in \mathscr{F}^n$,

$$P(x(t_n) \in B \,|\, x(t_1), x(t_2), \ldots, x(t_{n-1})) = P(x(t_n) \in B \,|\, x(t_{n-1})). \quad (1.5.2)$$

We give various equivalent formulations of the definition of the Markov process.

**Theorem 1.5.2.**   Each of the following conditions is equivalent to relation (1.5.2):

(i)   For $t_0 \leq s \leq t$, $t_0, s, t \in I$, and $B \in \mathscr{F}^n$,

$$P(x(t) \in B|\mathscr{F}_s) = P(x(t) \in B|x(s)),$$

where $\mathscr{F}_s \equiv \mathscr{F}[t_0, s]$ is the smallest sub-$\sigma$-algebra of $\mathscr{F}$ generated by all random variables $x(u)$ for $t_0 \leq u \leq s$.

(ii)   For $t_0 \leq s \leq t \leq \alpha$, $t_0, s, t \in I$, and $y$ $\mathscr{F}_a$-measurable and integrable,

$$E(y|\mathscr{F}_s) = E(y|x(s)).$$

(iii)   For $t_0 \leq s \leq t \leq a$ and $A \in \mathscr{F}_a$,

$$P(A|\mathscr{F}_s) = P(A|x(s)).$$

(iv)   For $t_0 \leq t_1 \leq t \leq t_2 \leq a$, $A_1 \in \mathscr{F}_t$, $A_2 \in \mathscr{F}_a$,

$$P(A_1 \cap A_2|\mathscr{F}_t) = P(A_1|x(t))P(A_2|x(t)).$$

As indicated in Section 1.4, for the conditional probability $P(x(t) \in B|x(s))$, there exists a conditional distribution $P(x(t) \in B|x(s)) = P(s, x(s), t, B)$ that is a function of four arguments $s, t \in I$ with $s \leq t$, $x \in R^n$, and $B \in \mathscr{F}^n$. This function has the following properties:

(a)   For fixed $s \leq t$ and $B \in \mathscr{F}^n$, we have

$$P(s, x(s), t, B) = P(x(t) \in B|x(s)) \quad \text{w.p. 1.}$$

(b)   $P(s, x, t, \cdot)$ is a probability on $\mathscr{F}^n$ for fixed $s \leq t$ and $x \in R^n$.

(c)   $P(s, \cdot, t, B)$ is $\mathscr{F}^n$-measurable for fixed $s \leq t$ and $B \in \mathscr{F}^n$.

(d)   For $t_0 \leq s \leq u \leq t \leq a$ and $B \in \mathscr{F}^n$ and for all $x \in R^n$, we have the Chapman–Kolmogorov equation

$$P(s, x, t, B) = \int_{R^n} P(u, y, t, B)P(s, x, u, dy) \quad \text{w.p. 1.} \tag{1.5.3}$$

**Definition 1.5.8.**   A function $P(s, x, t, B)$ with the above properties (b), (c), and (d) is called a transition probability or transition function. If $x(t)$ is a Markov process and $P(s, x, t, B)$ is a transition probability of the Markov process such that property (a) is satisfied, then $P(s, x, t, B)$ is called a transition probability of the Markov process $x(t)$. We shall also denote $P(s, x, t, B)$ by $P(x(t) \in B|x(s) = x)$.

If the probability $P(s, x, t, \cdot)$ has a density, that is, if for all $t, s \in I$, $s < t$, all $x \in R^n$ and all $B \in \mathscr{F}^n$, we have

$$P(s, x, t, B) = \int_B p(s, x, t, y) \, dy.$$

**Definition 1.5.9.** A Markov process $x(t)$ is said to be homogeneous if its transition probability $P(s, x, t, B)$ is stationary, that is, if the function

$$P(s + u, x, t + u, B) = P(s, x, t, B).$$

Let us consider a Markov process $x(t)$, $t \in I$ defined on $(\Omega, \mathcal{F} P)$ with state space $(R^n, \mathcal{F}^n)$ and transition function $P$. A random time $\tau$ is an $\mathcal{F}$-measurable mapping of $\Omega$ into $[0, a]$ such that $P(\tau < a) > 0$. A Markov time is a random time $\tau$ such that for all $t \in I$, the set $\{\omega : \tau(\omega) \leq t\} \in \mathcal{F}_t$, the $\sigma$-algebra generated by $x(s)$, $t_0 \leq s \leq t$. Let $\Omega^\tau = \{\omega : \tau(\omega) \leq a\}$, $\mathcal{F}^\tau = (\Omega^\tau \cap A : A \in \mathcal{F})$, $P^\tau = P/P(\Omega^\tau)$. Denote $\mathcal{F}_\tau$ the sub-$\sigma$-algebra of $\mathcal{F}^\tau$ such that $A \in \mathcal{F}^\tau$ implies $A \cap \{\omega : \tau < t\} \in \mathcal{F}_t$ for all $t \in I$.

**Definition 1.5.10.** A Markov process $x(t)$, $t \in I$, is said to be a jump Markov process if its transition function satisfies the relations

$$(a) \quad \lim_{t \to s^+} \frac{P(s, x, t, B) - I_B(x)}{t - s} = q(s, x, B)$$

uniformly in $(s, x, B)$, where $t \in I$, $x \in R^n$, and $B \in \mathcal{F}$.

(b) For fixed $(x, B)$, the function $q(s, x, B)$ is continuous with respect to $s \in I$ and uniformly continuous with respect to $(x, B)$.

**Remark 1.5.1.** It follows from (a) that there exists a $K > 0$ such that

$$|q(s, x, B)| \leq K, \qquad s \in I, \quad x \in R^n, \quad B \in \mathcal{F}.$$

**Example 1.5.1.** A Weiner process is a homogeneous $n$-dimensional Markov process $\omega(t)$ defined on $[0, \infty)$ with stationary transition probability

$$P(t, x, \cdot) = \begin{cases} \sqrt{\dfrac{1}{2\pi t}} \exp\left[-\dfrac{1}{2t} x^T x\right], & t > 0 \\ \text{Probability measure centered at } x, & t = 0. \end{cases}$$

An important class of random processes is the class of stationary processes. These are processes whose probabilistic characteristics do not change with displacement of time. More precisely, we define as follows:

**Definition 1.5.11.** A random process $x \in R[I, R[\Omega, R^n]]$ is said to be strictly stationary if for arbitrary $m, h$, and $t_1, t_2, \ldots, t_m$ such that $t_j + h \in I$, $i = 1, 2, \ldots, m$, the joint distribution function of random vectors $x(t_1 + h)$, $x(t_2 + h), x(t + h), \ldots, x(t_m + h)$ is independent of $h$, i.e.,

$$F_{t_1 + h, \, t_2 + h, \, \ldots, t_m + h}(x_1, x_2, \ldots, x_m) = F_{t_1, t_2, \ldots, t_m}(x_1, x_2, \ldots, x_m),$$

where $x_i \in R^n$, $i = 1, 2, \ldots, m$.

**Definition 1.5.12.** A random process $x \in R[I, R[\Omega, R^n]]$ is said to be a process with stationary increment if the joint distribution of the differences $x(t_2 + h) - x(t_1 + h)$, $x(t_3 + h) - x(t_2 + h)$, $\ldots$, $x(t_m + h) - x(t_{m-1} + h)$ is independent of $h$ for arbitrary $m, h$ and for $t_1, t_2, \ldots, t_m$ such that $t_i + h \in I$, $i = 1, 2, \ldots, m$.

**Remark 1.5.2.** The definition of a stationary process is equivalent to the following: For any bounded continuous function $f: R^n \to R$, the mean $E[f(x(t_1 + h), x(t_2 + h), \ldots, x(t_m + h))]$ is independent of $h$ for arbitrary $m, h$, and for $t_1, t_2, \ldots, t_m$ such that $t_i + h \in I$ for $i = 1, 2, \ldots, m$.

**Remark 1.5.3.** For every continuous function $f: R^n \to R$ and a stationary process $x(t)$, the random process

$$y(h) = f(x(t_1 + h), x(t_2 + h), \ldots, x(t_m + h))$$

is also stationary.

**Definition 1.5.13.** A process $x \in R[I, R[\Omega, R^n]]$ is said to have uncorrelated increments if

$$E[\|x(t) - x(s)\|^2] < \infty, \qquad t, s \in I, \tag{1.5.4}$$

and if whenever $s_1 \leq t_1 \leq s_2 \leq t_2$, the increments $x(t_1) - x(s_1)$ and $x(t_2) - x(s_2)$ are uncorrelated with each other, i.e.,

$$E\{(x(t_2) - x(s_2))^T(x(t_1) - x(s_1))\} = E(x(t_2) - x(s_2))^T E(x(t_1) - x(s_1)), \tag{1.5.5}$$

where $x$ is a column vector.

**Definition 1.5.14.** A process $x \in R[I, R[\Omega, R^n]]$ is said to have orthogonal increments if (1.5.4) holds and (1.5.5) is replaced by

$$E\{(x(t_2) - x(s_2))^T(x(t_1) - x(s_1))\} = 0.$$

**Remark 1.5.4.** If the process $x(t)$ has uncorrelated increments, then the process $y(t)$ defined by

$$y(t) = x(t) - E(x(t))$$

has uncorrelated and orthogonal increments.

**Remark 1.5.5.** If a process has independent increments and satisfies (1.5.5), then the process has uncorrelated increments.

**Remark 1.5.6.** Let $x \in R[I, R[\Omega, R^n]]$ be a process with orthogonal increments. Then $F(t)$ can be defined to satisfy

$$E[\|x(t) - x(s)\|^2] = F(t) - F(s), \qquad s < t. \tag{1.5.6}$$

Note that $F(t)$ is monotone nondecreasing and is determined by (1.5.6) up to an additive constant. For example, one can define

$$F(t) = \begin{cases} E[\|x(t) - x(t_0)\|^2], & t \geq t_0, \\ -E[\|x(t) - x(t_0)\|^2], & t < t_0. \end{cases}$$

## 1.6.  SEPARABILITY OF RANDOM PROCESSES

It is obvious from the properties of random functions that we cannot determine the probability of events defined by means of an uncountable set of index values. For example, the set $\{\omega : x(t, \omega) \geq 0 \text{ for all } t \in I\} = \bigcap_{t \in I} \{\omega : x(t, \omega) \geq 0\}$ may not be an event since it involves an uncountable intersection of events. Similarly, $y = \sup_{t \in I} x(t, \omega)$ may not be a random variable because sets of the form

$$\{\omega : y(\omega) \leq y\} = \bigcap_{t \in I} \{\omega : x(t, \omega) \leq y\}$$

may not be events. The satisfactory removal of this difficulty is due to J. L. Doob who introduced the notion of separability.

**Definition 1.6.1.**   A process $x \in R[I, R[\Omega, R^n]]$ is said to be separable if there exist a countable set $S \subset I$ and a fixed null event $\Lambda$ such that for any closed set $F \subset R^n$ and any open interval $O$, the two sets

$$\{\omega : x(t, \omega) \in F, t \in I \cap O\}, \qquad \{\omega : x(t, \omega) \in F, t \in I \cap S\}$$

differ by a subset of $\Lambda$. The countable set $S$ is called the separant or separating set.

The usefulness of the separability concept is demonstrated by the following.

**Theorem 1.6.1.**   The following properties are equivalent to the separability of process $x \in R(I, R[\Omega, R^n])$ with the separating set $S$: for every open interval $0 \subset I$,

(i)   $\inf_{t \in S \cap 0} x(t) = \inf_{t \in 0 \cap I} x(t), \sup_{t \in S \cap 0} x(t) = \sup_{t \in 0 \cap I} x(t);$

(ii)  $\inf_{t \in S \cap 0} x(t) \leq \inf_{t \in 0 \cap I} x(t), \sup_{t \in 0 \cap I} x(t) \geq \sup_{t \in S \cap 0} x(t);$

(iii) $\inf_{t \in S \cap 0} x(t) \leq x(t) \leq \sup_{t \in S \cap 0} x(t);$

(iv)  $\liminf_{t_n \to t_0} x(t_n) = \liminf_{t \to t_0} x(t), \limsup_{t_n \to t_0} x(t_n) = \limsup_{t \to t_0} x(t);$

(v)   $\liminf_{t_n \to t_0} x(t_n) \leq x(t) \leq \limsup_{t_n \to t_0} x(t_n);$

(vi)  $\liminf_{t_n \to t_0} x(t_n) \leq \liminf_{t \to t_0} x(t), \limsup_{t_n \to t_0} x(t_n) \leq \limsup_{t \to t_0} x(t);$

where $t_n \in S \cap 0$, $t_0 \in O \cap I$.

The next result shows that the separability requirement is not a serious restriction.

**Theorem 1.6.2.** For every stochastic process $x \in R[I, R[\Omega, R^n]]$ there exists a process $\bar{x} \in R[I, R[\Omega, R^n]]$ defined on the same probability space with values in $R^n$ such that

(i) $x(t)$ is separable,
(ii) $P\{x(t) = \bar{x}(t)\} = 1$ for each $t \in I$.

**Theorem 1.6.3.** Let $x(t)$ be a separable jump Markov process. Then

$$P(x(u) = x \text{ for all } u \in [s, t] | x(s) = x) = \exp\left[ -\int_s^t g(u, x)\, du \right].$$

## 1.7. DETERMINISTIC COMPARISON THEOREMS

In our later discussion, we shall need to employ deterministic comparison results several times. We shall give below some important results in this direction. Let $E$ be an open $(t, u)$-set in $R^{n+1}$. We shall mean by $C[E, R^n]$ the class of continuous functions from $E$ into $R^n$. We shall be using vectorial inequalities freely with the understanding that the same inequalities hold between their corresponding components.

We shall consider the deterministic differential system with an initial condition, written in the vectorial form

$$u' = g(t, u), \qquad u(t_0) = u_0, \tag{1.7.1}$$

where $g \in C[E, R^n]$.

We require that the function $g(t, u)$ satisfy certain monotonic properties.

**Definition 1.7.1.** The function $g(t, u)$ is said to possess a *quasi-monotone nondecreasing property* if for $u, v \in R^n$ such that $u \leq v$ and $u_i = v_i$, then $g_i(t, u) \leq g_i(t, v)$ for any $i = 1, 2, \ldots, n$ and fixed $t$.

We now quote the following deterministic comparison theorem [59]:

**Theorem 1.7.1.** Assume that

(i) $g \in C[E, R^n]$ and that $g(t, u)$ is quasi-monotone nondecreasing in $u$ for each $t$, where $E$ is an open $(t, u)$-set in $R^{n+1}$;
(ii) $[t_0, t_0 + a)$ is the largest interval of existence of the maximal solution $r(t) \equiv r(t, t_0, u_0)$ of (1.7.1);
(iii) $m \in C[[t_0, t_0 + a), R^n]$, $(t, m(t)) \in E$, $t \in [t_0, t_0 + a)$, and for a fixed Dini derivative $(D)$, the inequality

$$Dm(t) \leq g(t, m(t)) \tag{1.7.2}$$

holds for $t \in [t_0, t_0 + a)$. Then

$$m(t_0) \leq u_0 \tag{1.7.3}$$

implies

$$m(t) \leq r(t), \qquad t \in [t_0, t_0 + a]. \tag{1.7.4}$$

**Corollary 1.7.1.**   If in Theorem 1.7.1 inequalities (1.7.2) and (1.7.3) are reversed, then conclusion (1.7.4) is to be replaced by

$$m(t) \geq \rho(t), \qquad t \in [t_0, t_0 + a],$$

where $\rho(t) \equiv \rho(t, t_0, u_0)$ is the minimal solution of (1.7.1).

We shall consider an important result concerning the systems of integral inequalities that are reducible to differential inequalities.

**Theorem 1.7.2.**   Let assumptions (i) and (ii) of Theorem 1.7.1 hold, except the quasi-monotone nondecreasing property of $g(t, u)$ in $u$ for fixed $t$ is replaced by the nondecreasing property of $g(t, u)$ in $u$ for fixed $t$. Let $m \in C[[t_0, t_0 + a), R^n]$, $(t, m(t)) \in E$, $t \in [t_0, t_0 + a)$, $m(t_0) \leq u_0$, and

$$m(t) \leq m(t_0) + \int_{t_0}^{t} g(s, m(s))\, ds, \qquad t \in [t_0, t_0 + a).$$

Then

$$m(t) \leq r(t), \qquad t \in [t_0, t_0 + a].$$

**Notes**

The preliminary material concerning various definitions, notions, and results listed in Sections 1.1 to 1.6 is based on the well-known books and papers, namely, Doob [14], Gikhman and Skorokhod [19], Loéve [61], Neveu [74], Wong [87], Arnold [1], Iosifescu and Tăutu [27], Prohorov [75], and Skorokhod [80]. The results of Section 1.7 are adapted from the book by Lakshmikantham and Leela [59].

# CHAPTER 2

# SAMPLE CALCULUS APPROACH

## 2.0. INTRODUCTION

The dynamics of several biological, physical, and social systems of $n$ interacting species is described by an initial-value problem of the type

$$x'(t, \omega) = f(t, x(t, \omega), \omega), \qquad x(t_0, \omega) = x_0(\omega), \tag{2.0.1}$$

where $x$ is an $n$-vector, the prime represents a probabilistic derivative, and $f$ is a random rate process. Randomness arises in a system due to different types of unforeseen exogenous factors. This chapter emphasizes the study of systems (2.0.1) through sample calculus. In Section 2.1 we present essentials of sample calculus or a.s. sample calculus. We begin by defining stochastic and sample continuity of random processes and present some well-known results about them. A result that gives a sufficient condition for the $D$-boundedness of a subset of a space of sample continuous functions is also given. Next we define the sample differentiability of random functions. Finally, we discuss the measurability of random processes and then outline sample integrals of Riemann- and Lebesgue-type and state some useful results. We prove in Section 2.2 the basic sample existence theorem of Carathéodory-type for (2.0.1) and then study the continuation of sample solutions. Section 2.3 is devoted to the fundamental results concerning systems of random differential inequalities. Sufficient conditions are given for the existence of sample maximal and minimal solutions of (2.0.1) in Section 2.4. Section 2.5 deals with the basic random comparison theorems that are useful in estimating the sample solutions. Uniqueness and continuous dependence on parameters of sample solutions of (2.0.1) are discussed in Section 2.6. Moreover, the sample continuity and differentiability of solutions with respect to the initial state are also considered. The method of variation of parameters for sample solutions of linear and nonlinear systems of the type (2.0.1) forms the content of Section 2.7. These formulas give an alternate technique for studying the qualitative behavior of sample solutions of stochastic systems under constantly acting random disturbances. By employing the concept of random vector Lyapunov functions and the theory of random differential inequalities, very general comparison results are developed in Section 2.8. In Section 2.9 we discuss the scope of the general comparison theorems and illustrate them by means of several results in terms of logarithmic norm of a random matrix processes. The results developed in this section are computationally attractive. Depending on the mode of convergence in the probabilistic analysis, several stability notions relative to the given solution of (2.0.1) are formulated in Section 2.10. Finally, in Sections 2.11, 2.12, and 2.13 sufficient conditions for stability in probability, stability with probability one, and stability in the $p$th mean of the trivial solution of (2.0.1) are presented in a systematic and unified way. Some of the stability conditions are expressed in the framework of laws of large numbers and the random rate functions of the system.

## 2.1.  SAMPLE CALCULUS

In this section we shall discuss sample calculus or a.s. sample calculus for random functions. Because of Theorem 1.6.2, we can suppose without loss of generality that all random processes in our considerations are separable processes. Let $\Omega \equiv (\Omega, \mathscr{F}, P)$ be a complete probability space, and let $R[[a,b],$ $R[\Omega, R^n]]$ stand for a collection of all $R^n$-valued separable random functions or processes defined on $[a, b]$ with a state space $(R^n, \mathscr{F}^n)$, $a, b \in R$.

**Definition 2.1.1.** A random process $x \in R[[a,b], R[\Omega, R^n]]$ is said to be *continuous in probability* (or *stochastically continuous*) at $t \in (a, b)$ if for any $\eta > 0$,

$$P\{\omega : \|x(t + h, \omega) - x(t, \omega)\| \geq \eta\} \to 0 \qquad \text{as} \quad h \to 0.$$

If a process is continuous in probability at every $t \in (a, b)$ and has one-sided continuity in probability at the end points $a$ and $b$, then it is said to be continuous in probability on $[a, b]$.

We state a result that shows that every countable dense set in $[a, b]$ is a separating set.

(i)   Let $x \in R[[a,b], R[\Omega, R^n]]$. If $x$ is continuous in probability on $[a, b]$, then every countable dense set in $[a, b]$ is a separating set.

In the following, we list some results concerning stochastic continuity of well-known random functions.

(ii)   A random function $x \in R[[a, b], R[\Omega, R^n]]$ with independent increments is stochastically continuous at $t_0 \in (a, b)$ if and only if its characteristic
function $Q_t$ is continuous with respect to $t$ at $t_0$.

(iii)   A Gaussian process $x \in R[[a, b], R[\Omega, R^n]]$ is stochastically continuous at $t \in (a, b)$ if and only if the mean $m(t)$ and the variance $V(t)$ are
continuous functions of $t \in (a, b)$.

(iv)   Let $P(\lambda(t))$ be a separable and stochastically continuous Poisson
process with $\lim_{t \to a} \lambda(t) = \infty$, then all its sample paths are monotone nondecreasing step functions with probability one and with jumps equal to 1.

(v)   A jump Markov process is stochastically continuous.

**Definition 2.1.2.**   A random process $x \in R[[a, b], R[\Omega, R^n]]$ is said to be

(i)   *almost-surely continuous* (*a.s. continuous*) at $t \in (a, b)$ if

$$P\left\{\omega : \lim_{h \to 0}[\|x(t + h, \omega) - x(t, \omega)\|] \neq 0\right\} = 0,$$

(ii)   *sample continuous* (*almost-surely sample continuous*) in $t \in (a, b)$ if

$$P\left\{\bigcup_{t \in (a,b)}\left\{\omega : \lim_{h \to 0}[\|x(t + h, \omega) - x(t, \omega)\|] \neq 0\right\}\right\} = 0.$$

If a process is a.s. (a.s. sample) continuous at every $t \in (a, b)$ and has
one-sided a.s. (a.s. sample) continuity at the end points $a$ and $b$, then it is
said to be a.s. (a.s. sample) continuous on $[a, b]$.

We note that a.s. continuity of a process on $[a, b]$ is not equivalent to
sample continuity. This is simply because

$$\Omega(t) = \left\{\omega : \lim_{h \to 0}[\|x(t + h, \omega) - x(t, \omega)\|] \neq 0\right\},$$

being a null event for every $t \in [a, b]$, does not imply that the uncountable
union $\bigcup_{t \in [a,b]} \Omega(t)$ is a null event. However, if a process is sample continuous
on $[a, b]$, then it is necessarily a.s. continuous on $[a, b]$.

**Definition 2.1.3.**   A random process $x \in R[[a, b], R[\Omega, R^n]]$ is said to be

(i)   *bounded in probability* (*or stochastically bounded*) on $[a, b]$ if for every
$\eta > 0$, there exists a positive number $K$ such that

$$P\{\omega : \|x(t, \omega)\| > K\} < \eta \qquad \text{for all} \quad t \in [a, b];$$

(ii)   *sample bounded* (*or a.s. sample bounded*) on $[a, b]$ if there exists a
positive number $K$ such that

$$P\{\omega : \|x(t, \omega)\| > K \text{ for all } t \in [a, b]\} = 0.$$

Random functions that are sample or stochastic continuous enjoy all elementary properties of deterministic continuous functions.

Without proof, we state a few results that ensure the sample continuity of a random process. Let us denote

$$\alpha(\varepsilon, \delta) = \text{ess inf}_{\omega \in \Omega} \left[ \sup_{a \leq s \leq t \leq s + \delta \leq b} P\{\omega : \|x(s, \omega) - x(t, \omega)\| \geq \varepsilon | \mathscr{F}_s\} \right],$$

where $\mathscr{F}_s$ is a $\sigma$-algebra generated by a random function $x(u)$, $a \leq u \leq s$, and $x \in R[[a, b], R[\Omega, R^n]]$.

(i)   If for any $\varepsilon > 0$, $\lim_{\delta \to 0} \alpha(\varepsilon, \delta)/\delta = 0$, then the process $x(t)$ is sample continuous.

(ii)   Kolmogorov's theorem: if there exist positive constants $\beta$, $\gamma$, and $M$ such that for any $\eta > 0$, $t_1, t_2 \in [a, b]$,

$$P\{\omega : \|x(t_1, \omega) - x(t_2, \omega)\| > \eta\} \leq M|t_1 - t_2|^{1 + \gamma}/\eta^{\beta}, \qquad (2.1.1)$$

then $x(t)$ is sample continuous on $[a, b]$.

**Example 2.1.1.**   Assume that $x \in R[[a, b], R[\Omega, R^n]]$ and that it satisfies

$$E[\|x(t_1) - x(t_2)\|^2] \leq M|t_1 - t_2|^{1 + \gamma}, \qquad (2.1.2)$$

where $M$ and $\gamma$ are positive numbers and $t_1, t_2 \in [a, b]$. Then $x(t)$ is sample continuous on $[a, b]$. Applying Chebyshev's inequality to the left-hand member of inequality (2.1.2), one can obtain the inequality (2.1.1) with $\beta = 2$.

**Remark 2.1.1.**   We note that $x \in R[[a, b], R[\Omega, R^n]]$ and that $x$ is a stochastically continuous Gaussian process. Then $x$ is a sample continuous process on $[a, b]$.

**Definition 2.1.4.**   A random process $x \in R[[a, b], R[\Omega, R^n]]$ is said not to have *discontinuities of the second kind* at $t \in (a, b)$ if the realizations of the process have left- and right-hand limits at $t \in (a, b)$ and have right-hand and left-hand limits at the left and right end points of $[a, b]$, respectively.

The following results describe the structure of the sample paths of a separable stochastically continuous process having no discontinuities of the second kind.

(i)   Let $x \in R[[a, b], R[\Omega, R^n]]$ be a stochastically continuous process having no discontinuities of the second kind. Then there exists a process $\bar{x}(t)$ equivalent to it, and it is right-hand sample continuous on $[a, b]$.

(ii)   Let $x \in R[[a, b], R[\Omega, R^n]]$. If for any $\varepsilon > 0$, $\lim_{\delta \to 0} \alpha(\varepsilon, \delta) = 0$, then $x(t)$ has no discontinuities of the second kind.

We remark that under the hypothesis of (ii), the process is stochastically continuous.

We now state some results concerning the discontinuities of the second kind of well-known stochastic processes.

(i)   Assume that $x \in R[[a, b], R[\Omega, R^n]]$. Then every stochastically continuous process $x$ with independent increments has no discontinuities of the second kind on $[a, b]$.

(ii)   Assume that $x \in R[[a, b], R[\Omega, R^n]]$ and $x$ is a stochastically continuous process with independent increments. Then $x$ is a sample right-continuous process on $[a, b]$.

(iii)   Let $x$ be a jump Markov process that belongs to $R[[a, b], R[\Omega, R^n]]$. Then it has no discontinuities of the second kind on $[a, b]$.

The following result gives a sufficient condition for the total $D$-boundedness of a set in $C[[a, b], R[\Omega, R^n]]$, where $C[[a, b], R[\Omega, R^n]]$ is a collection of all $R^n$-valued separable and sample continuous random functions defined on $[a, b]$ with a state space $(R^n, \mathscr{F}^n)$. It is obvious that $C[[a, b], R[\Omega, R^n]]$ is a separable complete metric space relative to sup norm.

**Lemma 2.1.1.**   Assume that

$$S = \{x \in C[[a, b], R[\Omega, R^n]] : E[\|x(a)\|^4] \leq C$$
$$\text{and } E[\|x(t) - x(s)\|^4] \leq \beta|t - s|^{3/2}, \, t, s \in [a, b]\}$$

for some positive numbers $C$ and $\beta$. Then the set $S$ is a totally $D$-bounded subset of $C[[a, b], R[\Omega, R^n]]$.

*Proof.*   The main idea of the proof is as follows: by applying Chebyshev's inequality and the Borel–Cantelli lemma, one can find for every $\varepsilon > 0$, $\gamma_1(\varepsilon)$ and $\gamma_2(\varepsilon)$ with

$$P\{\|x(a)\| > \gamma_1(\varepsilon)\} \leq \frac{\varepsilon}{2}$$

and

$$P\left\{ \sup_{a \leq t \leq s \leq b} \left[ \frac{\|x(t) - x(s)\|}{|t - s|^{1/16}} \right] > \gamma_2(\varepsilon) \right\} \leq \frac{\varepsilon}{2}.$$

Therefore,

$$P\left\{ \|x(a)\| > \gamma_1(\varepsilon) \text{ or } \sup_{a \leq t \leq s \leq b} \left[ \frac{\|x(t) - x(s)\|}{|t - s|^{1/16}} \right] > \gamma_2(\varepsilon) \right\} \leq \varepsilon,$$

which implies that

$$P\left\{ \|x(a)\| \leq \gamma_1(\varepsilon) \text{ or } \sup_{a \leq t \leq s \leq b} \left[ \frac{\|x(t) - x(s)\|}{|t - s|^{1/16}} \right] \leq \gamma_2(\varepsilon) \right\} > 1 - \varepsilon$$

for every $x \in S$. Setting

$$K_\varepsilon = \left\{ x \in C[[a,b], R[\Omega, R^n]] : \|x(a)\| \le \gamma_1(\varepsilon) \right.$$

$$\left. \text{and} \quad \sup_{a \le t \le s \le b} \left[ \frac{\|x(t) - x(s)\|}{|t - s|^{1/16}} \right] \le \gamma_2(\varepsilon) \right\},$$

it is clear that $P\{x \in K_\varepsilon\} > 1 - \varepsilon$. On the other hand, Ascoli–Arzela's theorem shows that $K_\varepsilon$ is a compact subset of $C[[a,b], R[\Omega, R^n]]$. In order to complete the proof of the lemma, it is enough to apply Prohorov's theorem, stated in Section 1.3.

We now study some basic ideas about the differential calculus of random functions. We first define the notion of Dini derivatives for random functions. Let $x \in C[[a,b], R[\Omega, R^n]]$. Set

$$\Delta x(t, h, \omega) = (1/h)[x(t+h, \omega) - x(t, \omega)]$$

for fixed $t \in [a,b]$, $t + h \in [a,b]$, and $h \ne 0$ an arbitrary small number. It is obvious that $\Delta x(t, h, \omega)$ is separable and sample continuous in $h$ for each fixed $t \in [a,b]$; therefore, by Theorem 1.6.1(iv), we conclude that

$$\limsup_{h_n \to 0^\pm} [\Delta x(t, h_n, \omega)] = \limsup_{h \to 0^\pm} [\Delta x(t, h, \omega)] \equiv D^\pm x(t, \omega),$$

and

$$\liminf_{h_n \to 0^\pm} [\Delta x(t, h_n, \omega)] = \liminf_{h \to 0^\pm} [\Delta x(t, h, \omega)] \equiv D_\pm x(t, \omega).$$

exist w.p. 1. The four generalized derivatives defined above are called Dini derivatives.

Let us now define the differentiability of a process.

**Definition 2.1.5.** A random function $x \in C[[a,b], R[\Omega, R^n]]$ is said to possess

(i)   *an almost-sure derivative (a.s. derivative)* $x'(t, \omega)$ at $t \in (a,b)$ if

$$P\left\{ \omega : \lim_{h \to 0} \left[ \left\| \frac{x(t+h, \omega) - x(t, \omega)}{h} - x'(t, \omega) \right\| \right] \ne 0 \right\} = 0,$$

(ii)  *a sample (a.s. sample) derivative* $x'(t, \omega)$ in $t \in (a,b)$ if

$$P\left\{ \bigcup_{t \in (a,b)} \left\{ \omega : \lim_{h \to 0} \left[ \left\| \frac{x(t+h, \omega) - x(t, \omega)}{h} - x'(t, \omega) \right\| \right] \ne 0 \right\} \right\} = 0.$$

If a process is differentiable at every $t \in (a,b)$ and it has one-sided derivatives at the end points $a$ and $b$, then it is said to be differentiable on $[a,b]$. We remark that sample differentiability of a process on $[a,b]$ implies a.s.

differentiability on $[a, b]$. Further, we note that the sample derivative is a random variable.

**Remark 2.1.2.**  The Wiener or Brownian process is sample continuous but nowhere differentiable on $[a, b]$.

It is often desirable to be able to define integrals of sample paths of random processes. Lebesgue integrability of almost all sample functions is clearly a necessity. Even if almost all sample functions of $x(t)$ are Lebesgue-integrable, the resulting integral $\int_a^b x(t, \omega)\, dt$ still may not be a random variable. Therefore, it is necessary to suppose that $x(t, \omega)$ is a jointly measurable function with respect to $\mathscr{F}^1 \times \mathscr{F}$, where $\mathscr{F}^1$ denotes the $\sigma$-algebra of Lebesgue-measurable sets in $[a, b]$.

**Definition 2.1.6.**  A random process $x \in R[[a, b], R[\Omega, R^n]]$ is said to be measurable (product-measurable) iff $x(t, \omega)$ is an $(\mathscr{F}^1 \times \mathscr{F})$-measurable function defined on $[a, b] \times \Omega$ with values in $R^n$.

The following result gives a sufficient condition to ensure the existence of a measurable and separable process for a given process.

**Theorem 2.1.1.**  Let $x \in R[[a, b], R[\Omega, R^n]]$ be a stochastically continuous process with state space $(R^n, \mathscr{F}^n)$ and Lebesgue-measurable index set $[a, b]$. Then there exists a process $\bar{x}(t)$ defined on the same probability space and state space such that

  (i)  $P\{x(t) = \bar{x}(t)\} = 1$ for each $t \in [a, b]$,
  (ii)  $\bar{x}(t)$ is separable, and
  (iii)  $\bar{x}(t)$ is measurable.

Let us denote by $M[[a, b], R[\Omega, R^n]]$ a collection of random functions in $R[[a, b], R[\Omega, R^n]]$ that are product-measurable on $([a, b] \times \Omega, \mathscr{F}^1 \times \mathscr{F}, m \times P)$, where $\Omega \equiv (\Omega, \mathscr{F}, P)$ and $([a, b], \mathscr{F}^1, m)$ are a complete probability space and a Lebesgue-measurable space, respectively.

The next result establishes the product measurability of a stochastic process.

**Lemma 2.1.2.**  Assume that $x \in R[[a, b], R[\Omega, R^n]]$ and it is a.s. sample continuous. Then $x(t, \omega)$ is product-measurable on $([a, b] \times \Omega, \mathscr{F}^1 \times \mathscr{F}, m \times P)$, where $(\Omega, \mathscr{F}, P)$ and $([a, b], \mathscr{F}^1, m)$ are a complete probability space and a Lebesgue-measurable space, respectively.

*Proof.*  We note that the function $\alpha : [0, 1] \to [a, b]$, defined by $\alpha(t) = a + (b - a)t$, is continuous, one-to-one, and onto, and that therefore $[0, 1]$ is homeomorphic to $[a, b]$. In the light of this, without loss of generality, we take $[0, 1]$ instead of $[a, b]$. Define partitions $\Delta_n : t_0 = 0 < t_1^n < \cdots < t_k^n < \cdots < t_n^n = 1$, where $t_k^n = k2^{-n}$ for $k = 0, 1, 2, \ldots, 2^n$. We construct a sequence of $([0, 1] \times \Omega)$-measurable functions as follows:

$$x_n(t, \omega) = x(t_k^n, \omega) \qquad \text{for} \quad t \in [t_k^n, t_{k+1}^n], \qquad \omega \in \Omega,$$

for $k = 0, 1, 2, \ldots, 2^n$. It is obvious that $x(t_k^n) \in R[\Omega, R^n]$ and the sequence $x_n(t, \omega)$ is product-measurable on $[0, 1] \times \Omega$. On the other hand,

$$\|x_n(t, \omega) - x(t, \omega)\| = \|x(t_k^n, \omega) - x(t, \omega)\| \qquad \text{for} \quad t \in [t_k^n, t_{k+1}^n].$$

Because of a.s. continuity, it then follows that for fixed $t$,

$$P\left\{\omega : \lim_{n \to \infty} [\|x_n(t, \omega) - x(t, \omega)\|] \neq 0\right\} = 0.$$

Thus $x(t, \omega)$ is the pointwise limit of the product-measurable sequence $x_n(t, \omega)$. Hence $x(t, \omega)$ is product-measurable.

We shall now introduce the integration of stochastic functions.

**Definition 2.1.7.** A random process $x \in C[[a, b], R[\Omega, R^n]]$ is said to be *sample Riemann-integrable* if for any partition $\Delta : a = t_0 < t_1 < \cdots < t_n = b$ and $t_{i-1} \leq s_i \leq t_i$, the Riemann sum of $x(t, \omega)$

$$S_n(\omega) = \sum_{i=1}^{n} x(s_i, \omega)(t_i - t_{i-1})$$

has limit in the a.s. sample sense as $\delta_n \to 0$, where $\delta_n = \max_{1 \leq i \leq n}\{t_i - t_{i-1}\}$ and is denoted by $\int_a^b x(s, \omega)\, dt$.

We note that this integral is a random variable.

**Definition 2.1.8.** A random process $x \in M[[a, b], R[\Omega, R^n]]$ is said to be *sample Lebesgue-integrable* if $x(t, \omega)$ is Lebesgue-integrable on $[a, b]$ w.p. 1, and it is also denoted by

$$\int_a^b x(t, \omega)\, dt.$$

We note that every nonnegative sample path can be approximated by a sequence of step functions. In the light of this, it is obvious that the sample Lebesgue integral is a random variable. Also, if $x \in M[[a, b], R[\Omega, R^n]]$, $A$ is a Lebesgue-measurable subset of $[a, b]$, and

$$\int_A E[\|x(t, \omega)\|]\, dt < \infty,$$

then $x(t, \omega)$ is sample Lebesgue-integrable over $A$, and

$$\int_A x(t, \omega)\, dt$$

defines a measurable random process.

We remark that these integrals satisfy all the properties of the corresponding deterministic integrals.

Finally we state a result which gives sufficient conditions for an integral
of a random process $x \in M[[a, b], R[\Omega, R]]$ to be bounded w.p. 1 or a.s.
sample bounded.

**Theorem 2.1.2** (De La Vallée–Poussin's theorem).   Let $x \in M[[a, b],$
$R[\Omega, R]]$. If there exists a positive increasing deterministic function $\Phi(r)$,
defined for $r \geq 0$, such that $\Phi(r) \to \infty$ as $r \to \infty$ and

$$\int_a^b |x(t, \omega)| \Phi(|x(t, \omega)|) \, dt < L \quad \text{w.p. 1,}$$

where the integral is a sample Lebesgue integral and $L$ is a positive constant
independent of $\omega \in \Omega$, then the random function $x(t)$ is sample Lebesgue-
integrable on $[a, b]$, and its integral possesses the following property: given
$\varepsilon > 0$, there exists a $\delta = \delta(\varepsilon)$ such that the relations $[\alpha, \beta] \subset [a, b], \beta - \alpha < \delta$
imply

$$\left| \int_\alpha^\beta x(t, \omega) \, dt \right| < \varepsilon \quad \text{w.p. 1}$$

The following theorem enables us to interchange the order of integration
and to calculate integrals with respect to product measures by iteration.

**Theorem 2.1.3** (Fubini's theorem).   Let $(\Omega_1, \mathscr{F}_1, \mu_1)$ and $(\Omega_2, \mathscr{F}_2, \mu_2)$ be
two complete measurable spaces and $f$ be an integrable function on
$(\Omega_1 \times \Omega_2, \mathscr{F}_1 \times \mathscr{F}_2, \mu_1 \times \mu_2)$. Then

(i)   for almost all $\omega_1$, the function $f(\omega_1, \omega_2)$ is an integrable function on
$\Omega_2$;
(ii)   for almost all $\omega_2$, the function $f(\omega_1, \omega_2)$ is an integrable function on
$\Omega_1$;
(iii)   $\int_{\Omega_2} f(\omega_1, \omega_2) \, d\mu_2$ is an integrable function on $\Omega_1$;
(iv)   $\int_{\Omega_1} f(\omega_1, \omega_2) \, d\mu_1$ is an integrable function on $\Omega_2$;
(v)   $\int_{\Omega_1} [\int_{\Omega_2} f(\omega_1, \omega_2) \, d\mu_2] \, d\mu_1 = \int_{\Omega_1 \times \Omega_2} f(\omega_1, \omega_2) \, d(\mu_1 \times \mu_2) =$
$\int_{\Omega_2} [\int_{\Omega_1} f(\omega_1, \omega_2) \, d\mu_1] \, d\mu_2$.

A useful complement of Fubini's theorem is Tonelli's theorem, which we
list below.

**Theorem 2.1.4** (Tonelli's theorem).   Let $(\Omega_1 \times \Omega_2, \mathscr{F}_1 \times \mathscr{F}_2, \mu_1 \times \mu_2) =$
$(\Omega_1, \mathscr{F}_1, \mu_1) \times (\Omega_2, \mathscr{F}_2, \mu_2)$ be the product of two $\sigma$-finite measure spaces. Let
$f$ be a positive $(\mu_1 \times \mu_2)$-measurable function. Then

(i)   for almost all $\omega_1$, the function $f(\omega_1, \omega_2)$ is $\mu_2$-measurable;
(ii)   the function (in the extended real sense) $\int_{\Omega_2} f(\omega_1, \omega_2) \, d\mu_2$ is $\mu_1$-
measurable;

(iii) $\int_{\Omega_1} \left[ \int_{\Omega_2} f(\omega_1, \omega_2) \, d\mu_2 \right] d\mu_1 = \int_{\Omega_1 \times \Omega_2} f(\omega_1, \omega_2) \, d(\mu_1 \times \mu_2)$ irrespective of whether the integrals have finite or infinite values.

**Theorem 2.1.5** (The strong law of large numbers). Let $\{x(t) : t \in R\}$ be a measurable strictly stationary random process with $E\{\|x_0\|\} < \infty$, and let $\mathscr{F}_B$ be the Borel field of invariant $\omega$-sets. Then

$$\lim_{t \to \infty} \frac{1}{t} \int_0^t x(s, \omega) \, ds = E\{x_0 | \mathscr{F}_B\} \tag{2.1.3}$$

w.p. 1. In particular, if the process is metrically transitive, relation (2.1.3) reduces to

$$\lim_{t \to \infty} \frac{1}{t} \int_0^t x(s, \omega) \, ds = E\{x_0\}.$$

**Remark 2.1.3.** Note that the content of Theorem 2.1.5 is unchanged if the mean on the left in (2.1.3) is replaced by

$$\frac{1}{t} \int_u^{u+t} x(s, \omega) \, ds$$

**Definition 2.1.9.** A stationary process $\{x(t) : t \in R\}$ is said to be ergodic if it possesses invariant random variables.

## 2.2. EXISTENCE AND CONTINUATION

Let $R^n$ denote the $n$-dimensional Euclidean space with a convenient norm $\|\cdot\|$. We shall also use the same symbol for the corresponding norm of a matrix. As usual, let us use $R$ instead of $R^1$. Denote by $R_+$ the nonnegative real line. For some positive real number $\rho$, let

$$B(z, \rho) = [x \in R^n : \|z - x\| < \rho]$$

for a fixed $z \in R^n$. We denote by $\bar{B}(z, \rho)$ the closure of $B(z, \rho)$. Let $\Omega \equiv (\Omega, \mathscr{F}, P)$ be a complete probability space. We employ the following notation:

$C[R_+ \times B(z, \rho), R[\Omega, R^n]]$ is the class of sample continuous $R^n$-valued random functions $f(t, x)$ whose realizations are denoted by $f(t, x, \omega)$.

$M[R_+ \times B(z, \rho), R[\Omega, R^n]]$ is the class of $R^n$-valued random functions $f(t, x)$ such that $f(t, x(t))$ is product-measurable whenever $x(t)$ is product-measurable.

$IB[I, R[\Omega, R_+]]$ is the class of $K \in M[I, R[\Omega, R_+]]$ whose sample Lebesgue integral is bounded w.p. 1.

If $x$ is a random variable, then the realization of the random function $f(t, x)$ is denoted by $f(t, x(\omega), \omega)$. We shall assume that all relations such as equations, inequalities, belonging to, and limits involving random variables or functions are valid w.p. 1. Sometimes we shall interchangeably denote the random function by $x(t)$ or $x(t, \omega)$.

Let us consider a system of first-order random differential equations

$$x' = f(t, x, \omega), \qquad x(t_0, \omega) = x_0(\omega), \qquad (2.2.1)$$

where the prime denotes the sample derivative of $x$ and $f \in M[R_+ \times B(z, \rho), R[\Omega, R^n]]$.

**Definition 2.2.1.**   A random process $x(t)$ is said to be a sample solution process or a sample solution of the initial-value problem (2.2.1) on an interval $J = [t_0, t_0 + a]$ if it satisfies the following conditions:

  (i)   $x(t_0) = x_0$,
  (ii)  $x(t)$ is sample continuous,
  (iii) $x(t)$ is product measurable,
  (iv)  $x'(t, \omega) = f(t, x(t, \omega), \omega)$ w.p. 1 for almost every $t \in J$.

It is clear by (ii) and (iv) that $x(t)$ is an absolutely sample continuous and separable random process.

The proposed method for establishing the existence of solutions of (2.2.1) depends on its equivalent integral equation. This is accomplished by the following lemma.

**Lemma 2.2.1.**   Assume   that   $f \in M[J \times \bar{B}(z, \rho), R[\Omega, R^n]]$   and   that $f(t, x(t), \omega)$ is integrable on $J$ whenever $x(t)$ is a sample continuous random process. Then the initial-value problem (2.2.1) is equivalent to the integral equation

$$x(t, \omega) = x_0(\omega) + \int_{t_0}^{t} f(s, x(s, \omega), \omega) \, ds, \qquad (2.2.2)$$

where the integral denotes the a.s. sample Lebesgue integral.

The proof of this lemma is a consequence of the properties of the derivative of the indefinite integral of an integrable function and an absolutely continuous function.

We shall now prove the basic existence theorem of Carathéodory-type for (2.2.1).

**Theorem 2.2.1.**   Assume that

  (i)   $f \in M[J \times \bar{B}(z, \Omega), R[\Omega, R^n]]$ and $f(t, x, \omega)$ is sample continuous in $x$ for each $t \in J$;

(ii)   $K \in IB[J, R[\Omega, R_+]]$ and satisfies

$$\|f(t, x, \omega)\| \le K(t, \omega) \qquad \text{for} \quad (t, x) \in J \times \bar{B}(z, \rho); \qquad (2.2.3)$$

(iii)   $x_0 \in \bar{B}(z, \rho/2)$, $\rho$ being a positive number.

Then the initial-value problem (2.2.1) possesses at least one solution

$$x(t) = x(t, t_0, x_0) \qquad \text{on} \quad [t_0, t_0 + b] \quad \text{for some} \quad b > 0.$$

*Proof.*   We set

$$M(t, \omega) = \int_{t_0}^{t} K(s, \omega) \, ds \qquad \text{for} \quad t \in J \qquad (2.2.4)$$

and note that $M(t, \omega)$ is absolutely sample continuous with respect to $t$. By Tonelli's theorem, it follows that $M(t, \omega)$ is a random variable for fixed $t \in J$. Hence by Lemma 2.1.2, $M(t, \omega)$ is product-measurable. Moreover, because of the nonnegativity of $K(t, \omega)$, $M(t, \omega)$ is nondecreasing in $t$ w.p. 1. Also $M(t_0, \omega) = 0$, and $M(t, \omega)$ is bounded w.p. 1. We can therefore find a positive number $b \le a$ such that for $t \in [t_0, t_0 + b]$, $M(t, \omega) < \rho/2$ w.p. 1.

Let us define a sequence of random functions $x_m(t)$ on $[t_0, t_0 + b]$ by setting

$$x_m(t, \omega) = \begin{cases} x_0(\omega) & \text{for} \quad t_0 \le t \le t_0 + \dfrac{b}{m}, \\[2mm] x_0(\omega) + \displaystyle\int_{t_0}^{t - b/m} f(s, x_m(s, \omega), \omega) \, ds \\[2mm] & \text{for} \quad t_0 + \dfrac{ib}{m} \le t \le t_0 + \dfrac{b(i+1)}{m}, \end{cases} \qquad (2.2.5)$$

where $i = 1, 2, \ldots, m - 1$.

We shall first show that $x_m(t)$ is defined, product-measurable, and sample continuous on $[t_0, t_0 + b]$. Furthermore, we shall also show that $x_m(t) \in \bar{B}(z, \rho)$ w.p. 1.

For $t \in [t_0, t_0 + b/m]$, it is obvious from (2.2.5) and (iii) that $x_m(t)$ is defined, $x_m(t) \in \bar{B}(z, \rho)$, and $x_m(t)$ is a product-measurable and sample continuous random process. It then follows that $f(t, x_m(t), \omega)$ satisfies (2.2.3), and from (i) this implies that $f(t, x_m(t), \omega)$ is a product-measurable random function on $[t_0, t_0 + b/m]$. By the nature of $K(t, \omega)$, it is easy to see that

$$\int_{t_0}^{t} f(s, x_m(s), \omega) \, ds \qquad \cdot \qquad (2.2.6)$$

exists for $t \in [t_0, t_0 + b/m]$, where the integral represents the sample Legesgue integral. In view of (2.2.6) and (2.2.5), $x_m(t)$ is defined on $[t_0 + b/m, t_0 + 2b/m]$. Furthermore, it is clear that

$$x_m(t, \omega) = x_m\left(t_0 + \frac{b}{m}, \omega\right) + \int_{t_0 + b/m}^{t} f\left(s - \frac{b}{m}, x_m\left(s - \frac{b}{m}, \omega\right), \omega\right) ds \qquad (2.2.7)$$

exists on $[t_0 + b/m, t_0 + 2b/m]$. Note that for $t \in [t_0 + b/m, t_0 + 2b/m]$, $f(t - b/m, x_m(t - b/m, \omega), \omega)$ is equal to $f(t, x_m(t, \omega), \omega)$ on $[t_0, t_0 + b/m]$. This implies that $x_m(t)$ is sample continuous on $[t_0 + b/m, t_0 + 2b/m]$. Hence by the product measurability of $f(t, x_m(t), \omega)$ on $[t_0, t_0 + b/m]$ and Tonelli's theorem, it follows that $x_m(t)$ is a random variable for a fixed $t \in [t_0 + b/m, t_0 + 2b/m]$. By Lemma 2.1.2, $x_m(t)$ is product-measurable. Because of the fact that

$$x_m(t) \in \bar{B}(z, \rho) \qquad \text{for} \quad t \in [t_0, t_0 + b/m],$$

relations (2.2.5), (2.2.3), and (2.2.4) and the choice of $b$ assure that

$$\|x_m(t) - x_0\| \le M(t - b/m, \omega) \qquad \text{for} \quad t \in [t_0, t_0 + 2b/m], \quad (2.2.8)$$

which implies that

$$x_m(t) \in B(z, \rho) \qquad \text{for} \quad t \in [t_0 + b/m, t_0 + 2b/m].$$

Thus $x_m(t) \in \bar{B}(z, \rho)$ is a product-measurable random process and is sample continuous on $[t_0, t_0 + 2b/m]$.

Let us assume that $x_m(t) \in \bar{B}(z, \rho)$ is a product-measurable and sample continuous random process on $[t_0, t_0 + jb/m]$ for $1 < j < m$ and follow the foregoing argument. From (2.2.5) we have

$$x_m(t, \omega) = x_m\left(t_0 + \frac{jb}{m}, \omega\right) + \int_{t_0 + jb/m}^{t} f\left(s - \frac{jb}{m}, x_m\left(s - \frac{jb}{m}, \omega\right), \omega\right) ds \quad (2.2.9)$$

for

$$t \in [t_0 + jb/m, t_0 + (j + 1)b/m].$$

We note that for $t \in [t_0 + jb/m, t_0 + b(j + 1)/m]$,

$$f(t - jb/m, x(t - jb/m, \omega), \omega) = f(t, x_m(t, \omega), \omega)$$

for $t \in [t_0 + (j - 1)b/m, t_0 + jb/m]$.

Since $x_m(t)$ is product-measurable and sample continuous on $[t_0, t_0 + jb/m]$, so also it is product-measurable and sample continuous on $[t_0 + (j - 1)b/m, t_0 + jb/m]$. Hence $f(t, x_m(t), \omega)$ is product-measurable on $[t_0 + (j - 1)b/m, t_0 + jb/m]$, and by Tonelli's theorem, $x_m(t)$ in (2.2.9) is a random variable for fixed $t \in [t_0 + jb/m, t_0 + (j + 1)b/m]$. Since $x_m(t) \in \bar{B}(z, \rho)$ for $t \in [t_0, t_0 + jb/m]$, we get $x_m(t) \in \bar{B}(z, \rho)$ for $t \in [t_0 + (j - 1)/m, t_0 + jb/m]$.

This, together with the properties of $K(t, \omega)$ in (ii), implies that the indefinite Lebesgue sample integral in (2.2.9) exists, which shows that $x_m(t)$ is sample continuous on $[t_0 + jb/m, t_0 + (j + 1)b/m]$. Now by Lemma 2.1.2, $x_m(t)$ is product-measurable for $t \in [t_0 + jb/m, t_0 + (j + 1)b/m]$. In view of the fact that $x_m(t) \in \bar{B}(z, \rho)$ for $t \in [t_0 + (j - 1)b/m, t_0 + jb/m]$, the relations

(2.2.3)–(2.2.5) and the choice of $b$ yield

$$\|x_m(t) - x_0\| \leq M(t - jb/m, \omega) \qquad \text{for} \quad t \in [t_0, t_0 + (j+1)b/m],$$

which implies that $x_m(t) \in \bar{B}(z, \rho)$ for $t \in [t_0 + jb/m, t_0 + (j+1)b/m]$.

Thus $x_m(t) \in \bar{B}(z, \rho)$ is a sample continuous and product-measurable random process on $[t_0, t_0 + (j+1)b/m]$. Therefore, by induction, (2.2.5) defines $x_m(t)$ as a product-measurable and sample continuous sequence of random functions on $[t_0, t_0 + b]$ such that

$$x_m(t, \omega) = x_0(\omega) \qquad \text{w.p. 1 on} \quad [t_0, t_0 + b/m] \qquad (2.2.10)$$

and

$$\|x_m(t, \omega) - x_0(\omega)\| \leq M(t - b/m, \omega) \qquad \text{on} \quad [t_0 + ib/m, t_0 + (i+1)b/m] \tag{2.2.11}$$

for $i = 1, 2, \ldots . m - 1$. From (2.2.10), (2.2.11), sample boundedness of $M(t, \omega)$, and the choice of $b$, we conclude that the collection $\{x_m(t)\}$ is sample equibounded on $[t_0, t_0 + b]$. Now, we shall show that the collection $\{x_m(t)\}$ is sample uniformly continuous on $[t_0, t_0 + b]$. From (2.2.3) and (2.2.5), we have

$$\|x_m(t, \omega) - x_m(s, \omega)\| \leq M(t - b/m, \omega) - M(s - b/m, \omega)$$

$$\text{for} \quad t, s \in [t_0, t_0 + b].$$

Since $M(t, \omega)$ is sample continuous on $[t_0, t_0 + b]$, so also it is sample uniformly continuous on $[t_0, t_0 + b]$. This implies that the family $\{x_m(t)\}$ is sample equicontinuous on $[t_0, t_0 + b]$.

We recall that $C[[t_0, t_0 + b], R^n]$ is the collection of continuous functions defined on $[t_0, t_0 + b]$ with values in $R^n$. It is a separable complete metric space with the metric $d$ defined by

$$d(x, y) = \max_{t \in [t_0, t_0 + b]} \|x(t) - y(t)\| \qquad \text{for} \quad x, y \in C[[t_0, t_0 + b], R^n].$$

Let $R[\Omega, C[[t_0, t_0 + b], R^n]]$ be the system of $C[[t_0, t_0 + b] R^n]$-valued random variables. It is obvious that the collection $\{x_m(t)\}$ can be viewed as a $C[[t_0, t_0 + b], R^n]$-valued random variable. Hence $\{x_m\} \subseteq R[\Omega, C[[t_0, t_0 + b], R^n]]$. Recall that the family $\{x_m(t)\}$ is sample equicontinuous and equibounded, and hence by Ascoli–Arzela's theorem, $\{x_m(t)\}$ is a compact subset of $C[[t_0, t_0 + b], R^n]$. Recall that $x_m(t) \in \bar{B}(z, \rho)$ for all $t \in [t_0, t_0 + b]$ w.p. 1. Now by application of Prohorov's theorem, $\{x_m\}$ is totally $D$-bounded in $R[\Omega, C[[t_0, t_0 + b], R^n]]$. We also note that $\{x_m(t_0) \equiv x_0\}$ is compact. Hence $(x_m, x_0)$ is also compact because the direct product of a finite number of compact sets is compact if and only if each set is compact. Thus by the application of Prohorov's theorem, there exists a $D$-Cauchy subsequence

$\{(x_{m_r}, x_0)\}$ of $\{(x_m, y_0)\}$. Let us denote this sequence $x_{m_r}$ by $\{x_r\}$. By Skorokhod's theorem, we can construct a sequence $\{(Y_r, Y_{0_r})\}$ and $x$ such that

$$Y_r, x \in R[\Omega, C[[t_0, t_0 + b], R^n]],$$

$$D((Y_r, Y_{0r}), (x_r, x_0)) = 0, \qquad \text{for} \quad r = 1, 2, \ldots \tag{2.2.12}$$

and

$$P((Y_r, Y_{0r}) \to (x, x_0)) = 1. \tag{2.2.13}$$

It is obvious that $x(t)$ is a sample continuous and product-measurable random variable on $[t_0, t_0 + b]$. Notice that $D((Y_r, Y_{0r}), (x_r, x_0)) = 0$ means that $(Y_r, Y_{0r})$ and $(x_r, x_0)$ have the same distribution. Hence $Y_r \in \bar{B}(z, \rho)$ w.p. 1, so also $x \in \bar{B}(z, \rho)$ w.p. 1 in view of (2.2.13).

Next we shall show that $x(t)$ is a solution process of (2.2.1). Since from (2.2.3) we have

$$\|f(t, Y_r(t, \omega), \omega)\| \le K(t, \omega) \qquad \text{on} \quad t \in [t_0, t_0 + b]$$

and $f$ is sample continuous in $x$ w.p. 1 for fixed $t \in [t_0, t_0 + b]$, it follows that

$$f(t, Y_r(t, \omega), \omega) \to f(t, x(t, \omega), \omega) \qquad \text{w.p. 1 as} \quad r \to \infty$$

for every fixed $t \in [t_0, t_0 + b]$. By the Lebesgue dominated convergence theorem, we get

$$\int_{t_0}^t f(s, Y_r(s, \omega), \omega)\, ds \to \int_{t_0}^t f(s, x(s, \omega), \omega)\, ds \quad \text{w.p. 1} \tag{2.2.14}$$

as $r \to \infty$ for any $t \in [t_0, t_0 + b]$. Relations (2.2.5) and (2.2.12) and sample continuity of functions yield

$$Y_r(t, \omega) = x_0(\omega) + \int_{t_0}^t f(s, Y_r(s, \omega), \omega)\, ds - \int_{t-b/r}^t f(s, Y_r(s, \omega), \omega)\, ds. \tag{2.2.15}$$

By (2.2.3), (2.2.4), and the uniform sample continuity and boundedness of $M(t, \omega)$, it follows that the sample integral

$$\int_{t-b/r}^t f(s, Y_r(s, \omega), \omega)\, ds$$

tends to zero as $r \to \infty$. This, together with (2.2.12)–(2.2.15), shows, by letting $r \to \infty$, that

$$x(t, \omega) = x_0(\omega) + \int_{t_0}^t f(s, x(s, \omega), \omega)\, ds.$$

This completes the proof, by virtue of Lemma 2.2.1.

The following corollary of Theorem 2.2.1 is useful in applications.

**Corollary 2.2.1.** Let $D$ be an open set in $R^n$ and $E = J \times D$. Let $E_0$ be a compact subset of $E$. Suppose that $f \in M[E, R[\Omega, R^n]]$ and $f(t, x, \omega)$ is sample continuous in $x$ for each $t$. Let $K \in IB[J, R[\Omega, R_+]]$ and

$$\|f(t, x, \omega)\| \leq K(t, \omega) \qquad \text{for} \quad (t, x) \in E_0 \quad \text{w.p. 1.}$$

Then there exists a positive number $b$ such that if $(t_0, x_0) \in E_0$, (2.2.1) has a solution $x(t)$ on $[t_0, t_0 + b]$. Moreover, $x(t) \in E_0$ w.p. 1 for all $t \in [t_0, t_0 + b]$.

**Definition 2.2.2.** A solution process of (2.2.1) defined on an interval $J_1$ is said to be continuable if there exists a solution process $y(t)$ of (2.2.1) defined on an interval $J_2 \supset J_1$ such that $y(t) = x(t)$ w.p. 1 on $J_1$.

The next theorem deals with the problem of continuation or extention of solutions up to the boundary of $E$.

**Theorem 2.2.2.** Assume that $f \in M[E, R[\Omega, R^n]]$ and $f(t, x, \omega)$ is sample continuous in $x$ for each $t$. Furthermore, let $K \in IB[J, R[\Omega, R_+]]$ and

$$\|f(t, x, \omega)\| \leq K(t, \omega) \qquad \text{for} \quad (t, x) \in E_0 \quad \text{w.p. 1,}$$

where $\bar{E}_0$ is any compact subset of $E$. Let $x(t)$ be a solution process of (2.2.1) on some interval $[t_0, b_0]$ with $(b_0, \lim_{t \to \bar{b}_0} x(t)) \in E_0$ w.p. 1. Then the solution $x(t)$ can be continued to the right of $b_0$. Moreover, it can be continued to the boundary of $E$.

*Proof.* Since $(b_0, \lim_{t \to \bar{b}_0} x(t)) \in E_0$, by Corollary 2.2.1, there exists a solution $\hat{x}(t)$ through $(b_0, \lim_{t \to \bar{b}_0} x(t))$ on $[b_0, b_0 + b]$ for some $b > 0$. If $y(t)$ is defined by

$$y(t) = \begin{cases} x(t) & \text{for} \quad t \in [t_0, b_0], \\ \hat{x}(t) & \text{for} \quad t \in [b_0, b_0 + b], \end{cases}$$

then $y(t)$ is a solution on $[t_0, b_0 + b]$, and $y(t_0) = x_0$. The proof of the last part of the statement is similar to the proof of Theorem 1.1.3 in Lakshmikantham and Leela [59].

**Remark 2.2.1.** In the foregoing discussion we have assumed that $K \in IB[J, R[\Omega, R_+]]$. Instead of this, we can assume that $K$ satisfies the conditions of De La Vallée–Poussin's theorem.

We conclude this section by giving an example which shows that the type of general random differential equations considered include other important classes of equations.

**Example 2.2.1.** Let $\eta \in M[J, R[\Omega, R^m]]$. Let $F : J \times B(z, \rho_1) \times B(y, \rho_2) \to R^n$. Suppose that $f(t, x, y)$ is continuous in $(x, y)$ for fixed $t \in J$ and measurable in $t$ for fixed $(x, y)$. Set

$$f(t, x, \omega) = F(t, x, \eta(t, \omega)). \tag{2.2.16}$$

It is clear that if $f$ in (2.2.16) satisfies the hypotheses of Theorem 2.2.1, then the initial-value problem

$$x'(t, \omega) = f(t, x, \eta(t, \omega)), \qquad x(t_0, \omega) = x_0(\omega)$$

has at least one solution on $[t_0, t_0 + b]$.

We note that the random function $\eta(t, \omega)$ in Example 2.2.1 can be a separable and stochastically continuous process with independent increments (Poisson process, Gaussian process) or a separable jump Markov process.

## 2.3  RANDOM DIFFERENTIAL INEQUALITIES

We shall state and prove some fundamental results concerning systems of random differential inequalities.

**Theorem 2.3.1.**   Assume that

(H$_1$)   $g \in M[[t_0, t_0 + a) \times D, R[\Omega, R^n]]$, $g(t, u, \omega)$ is a.s. quasi-monotone nondecreasing in $u$ for each $t$, $D$ is an open set in $R^n$, and $a > 0$,

(H$_2$)   $u$, $v \in C[t_0, t_0 + a)$, $R[\Omega, R^n]]$ for $(t, u(t, \omega))$, $(t, v(t, \omega)) \in [t_0, t_0 + a) \times D$,

$$D_- v(t, \omega) \le g(t, v(t, \omega), \omega),$$

and

$$D_- u(t, \omega) > g(t, u(t, \omega), \omega) \qquad \text{a.e. in} \quad t \in [t_0, t_0 + a),$$

(H$_3$)   $v(t_0, \omega) < u(t_0, \omega)$.

Then

$$v(t, \omega) < u(t, \omega) \qquad \text{for} \quad t \in [t_0, t_0 + a). \tag{2.3.1}$$

*Proof.*   Suppose that (2.3.1) is false; then without loss of generality, there exist $t_1 > t_0$, an index $j$, and $\Omega_j \subset \Omega$ such that

(i)   $u_j(t_1, \omega) = v_j(t_1, \omega)$, $\omega \in \Omega_j$ with $P(\Omega_j) > 0$,
(ii)   $u_j(t, \omega) > v_j(t, \omega)$, $t \in [t_0, t_1)$ w.p. 1,
(iii)   $u_i(t_1, \omega) \ge v_i(t_1, \omega)$, $\omega \in \Omega_j$, $i \ne j$,
(iv)   hypothesis (H$_2$) holds for $(t_1, \omega)$, where $\omega \in (\Omega_j - N)$ and $N \subset \Omega_j$ is any null event.

For $\omega \in \Omega_j$ and small $h < 0$, we have

$$\frac{1}{h} [u_j(t_1 + h, \omega) - u_j(t_1, \omega)] < \frac{1}{h} [v_j(t_1 + h, \omega) - v_j(t_1, \omega)],$$

which implies that

$$D_-u_j(t_1, \omega) \le D_-v_j(t_1, \omega).$$

This, together with (H$_2$), for $\omega \in \Omega_j$, yields

$$g_j(t_1, u(t_1, \omega), \omega) < g_j(t_1, v(t_1, \omega), \omega). \tag{2.3.2}$$

On the other hand, from (H$_1$) and (i)–(iv), we have

$$g_j(t_1, v(t_1, \omega), \omega) \le g_j(t_1, u(t_1, \omega), \omega) \qquad \text{for} \quad \omega \in \Omega_j.$$

Inequality (2.3.2) now leads to a contradiction:

$$g_j(t_1, u(t_1, \omega), \omega) < g_j(t_1, u(t_1, \omega), \omega) \qquad \text{for} \quad \omega \in \Omega_j.$$

Hence $P(\Omega_j) = 0$ and such a $t_1$ does not exist. This proves the theorem.

**Remark 2.3.1.** It is obvious from the proof that the inequalities in hypothesis (H$_2$) can also be replaced by

$$D_-v(t, \omega) < g(t, v(t, \omega), \omega),$$

and

$$D_-u(t, \omega) \ge g(t, u(t, \omega), \omega), \qquad \text{a.e. in} \quad t \in [t_0, t_0 + a).$$

We note that the proof does not demand the validity of the inequalities in hypothesis (H$_2$) for all $(t, \omega) \in [t_0, t_0 + a) \times \Omega$. In fact, it is enough to assume that the inequalities

$$D_-v_j(t, \omega) \le g_j(t, v(t, \omega), \omega),$$
$$D_-u_j(t, \omega) > g_j(t, u(t, \omega), \omega)$$

are satisfied for $t \in \{t : u_j(t, \omega) = v_j(t, \omega)\}$ and $\omega \in \Omega_j(t) = \{\omega : u_j(t, \omega) = v_j(t, \omega)\}$ with $P(\Omega_j(t)) > 0$ and $1 \le j \le n$.

By assuming one-sided estimates on $g(t, u, \omega)$, hypothesis (H$_2$) can be relaxed, and a weaker kind of inequality can be established.

**Theorem 2.3.2.** Assume that hypotheses (H$_1$), (H$_2$) hold. Further assume that for $t \in [t_0, t_0 + a)$, $u$, $v$, and $g$ satisfy the inequalities

$$D_-v(t, \omega) \le g(t, v(t, \omega), \omega)$$

and

$$D_-u(t, \omega) \ge g(t, u(t, \omega), \omega) \qquad \text{a.e. in} \quad t \in [t_0, t_0 + a);$$

moreover, $g$ satisfies the one-sided estimate

$$g(t, x, \omega) - g(t, y, \omega) \le L(t, \omega)(x - y) \tag{2.3.3}$$

for $x, y \in D$ such that $x > y$, where $L \in M[[t_0, t_0 + a), R[\Omega, R_+^{n^2}]]$ and is sample Lebesgue-integrable on $[t_0, t_0 + a)$. Then $v(t_0, \omega) \le u(t_0, \omega)$ implies

$$v(t, \omega) \le u(t, \omega) \qquad \text{for} \quad t \in \lceil t_0, t_0 + a).$$

*Proof.* Set $w = u + \exp[2 \int_{t_0}^t L(s, \omega) ds]\varepsilon$, where $\varepsilon = (\varepsilon_1, \varepsilon_2, \dots, \varepsilon_n)^T$ and $\varepsilon_j > 0$ for all $1 \le j \le n$. It is obvious that

$$D_- w(t, \omega) = D_- u(t, \omega) + 2L(t, \omega) \exp\left[ 2 \int_{t_0}^t L(s, \omega) ds \right] \varepsilon$$

This, together with inequality (2.3.3), yields

$$D_- w(t, \omega) \ge g(t, u(t, \omega), \omega) + 2L(t, \omega) \exp\left[ 2 \int_{t_0}^t L(s, \omega) ds \right] \varepsilon$$

$$\ge g(t, w(t, \omega), \omega) + L(t, \omega) \exp\left[ 2 \int_{t_0}^t L(s, \omega) ds \right] \varepsilon$$

$$> g(t, w(t, \omega), \omega) \quad \text{w.p. } 1$$

since $w > u$. We note that $v(t_0, \omega) < \omega(t_0, \omega)$ w.p. 1. Now by Theorem 2.3.1, we have

$$v(t, \omega) < w(t, \omega) \qquad \text{for} \quad t \in [t_0, t_0 + a).$$

By taking the limit as $\varepsilon \to 0$, we get the desired inequality.

## 2.4. MAXIMAL AND MINIMAL SOLUTIONS

By introducing the notion of maximal and minimal solutions of

$$u'(t, \omega) = g(t, u(t, \omega), \omega), \qquad u(t_0, \omega) = u_0(\omega), \tag{2.4.1}$$

sufficient conditions are given for the existence of maximal and minimal solutions of (2.4.1).

**Definition 2.4.1.** Let $r(t, \omega)$ be a sample solution process of the system of random differential equations (2.4.1) on $[t_0, t_0 + a)$. Then $r(t, \omega)$ is said to be a sample maximal solution process of (2.4.1) if for every sample solution $u(t, \omega)$ existing on $[t_0, t_0 + a)$, the inequality

$$u(t, \omega) \le r(t, \omega), \qquad t \in [t_0, t_0 + a), \tag{2.4.2}$$

holds. A sample minimal solution $\rho(t, \omega)$ may be defined similarly by reversing inequality (2.4.2).

We shall now consider the existence theorem for maximal and minimal solution processes of (2.4.1).

**Theorem 2.4.1.** Assume that the hypotheses of Theorem 2.2.1 hold and that $g(t, u, \omega)$ possesses the sample quasi-monotone nondecreasing property in $u$ for fixed $t \in J$. Then there exist sample maximal and minimal solutions of (2.4.1) on $[t_0, t_0 + b]$ for some $b > 0$.

*Proof.* Let $\varepsilon > 0$ be such that $\|\varepsilon\| < \rho/4$. Consider the initial-value problem

$$u'(t, \omega) = g_\varepsilon(t, u(t, \omega), \omega), \qquad u(t_0, \omega) = u_0(\omega) + \varepsilon, \tag{2.4.3}$$

where $g_\varepsilon(t, u, \omega) = g(t, u, \omega) + \varepsilon$ and $u_0(\omega) \in \bar{B}(z, \rho/4)$. It is easy to see that $g_\varepsilon(t, u, \varepsilon)$ satisfies all the hypotheses of Theorem 2.2.1 with $K(t, \omega)$ replaced by $K(t, \omega) + \rho/4$. Hence by Theorem 2.2.1, the initial-value problem (2.4.3) has a sample solution $u(t, \omega, \varepsilon)$ for $t \in [t_0, t_0 + b]$ for some $b > 0$.

We shall prove the existence of the sample maximal solution only since the proof for the case of the sample minimal solution is similar.

Let $\varepsilon_1 < \varepsilon_2 \le \varepsilon$ and let $u_1(t, \omega) = u(t, \omega, \varepsilon_1), u_2(t, \omega) = u(t, \omega, \varepsilon_2)$ be solutions of

$$
\begin{aligned}
u_1'(t, \omega) &= g_{\varepsilon_1}(t, u_1(t, \omega), \omega), & u_1(t, \omega) &= u_0(\omega) + \varepsilon_1, \\
u_2'(t, \omega) &= g_{\varepsilon_2}(t, u_2(t, \omega), \omega), & u_2(t, \omega) &= u_0(\omega) + \varepsilon_2 \quad \text{w.p. 1.}
\end{aligned}
$$

This implies that

$$u_2'(t, \omega) > g_{\varepsilon_1}(t, u_2(t, \omega), \omega) \qquad \text{a.e. in} \quad t \in [t_0, t_0 + b]$$

and

$$u_2(t_0, \omega) > u_1(t_0, \omega).$$

By applying Theorem 2.3.1, we get

$$u_1(t, \omega) < u_2(t, \omega), \qquad t \in [t_0, t_0 + b].$$

Choose a decreasing sequence $\{\varepsilon_n\}$ such that $\varepsilon_n \to 0$ as $n \to \infty$. Thus the uniform limit

$$r(t, \omega) = \lim_{n \to \infty} u_n(t, \omega) \tag{2.4.4}$$

exists on $[t_0, t_0 + b]$ w.p. 1. It is obvious that $r(t_0, \omega) = u_0(\omega)$. Because of the continuity of $g(t, u, \omega)$ in $u$ for fixed $t \in [t_0, t_0 + b]$, the limit

$$\lim_{n \to \infty} g_{\varepsilon_n}(t, u_n(t, \omega), \omega) = g(t, r(t, \omega), \omega) \tag{2.4.5}$$

exists as $n \to \infty$ a.e. on $[t_0, t_0 + b]$. By the application of the Lebesgue dominated convergence theorem, we get

$$\lim_{n \to \infty} \int_{t_0}^{t} g_{\varepsilon_n}(s, u_n(s, \omega), \omega) \, ds = \int_{t_0}^{t} g(s, r(s, \omega), \omega) \, ds \quad \text{w.p. 1}$$

for $t \in [t_0, t_0 + b]$. This implies that $r(t, \omega)$ is a solution of (2.4.1).

We shall show that $r(t, \omega)$ is the desired maximal solution of (2.4.1). Let $u(t, \omega)$ be any solution of (2.4.1). Then by an application of Theorem 2.3.1, we have

$$u(t, \omega) < u_n(t, \omega) \quad \text{on} \quad [t_0, t_0 + b].$$

This proves the theorem since

$$u(t, \omega) \leq \lim_{n \to \infty} u_n(t, \varepsilon) \equiv r(t, \omega) \quad \text{w.p. 1}$$

uniformly on $[t_0, t_0 + b]$.

## 2.5. RANDOM COMPARISON PRINCIPLE

An important technique in the theory of random differential inequalities is concerned with estimating a random function satisfying a differential inequality by extremal solutions of a corresponding random differential equation. One of the results that has wide applicability is the following comparison theorem.

**Theorem 2.5.1.** Assume that

(H$_1$) $g \in M[E, R[\Omega, R^n]]$ and $g(t, u, \omega)$ is sample continuous in $u$ for fixed $t$, where $E = [t_0, t_0 + a) \times D$ and $D$ is an open set in $R^n$;

(H$_2$) $g(t, u, \omega)$ is a.s. quasi-monotone nondecreasing in $u$ for each fixed $t$;

(H$_3$) $r(t, \omega)$ is the sample maximal solution of the random differential system

$$u'(t, \omega) = g(t, u(t, \omega), \omega), \qquad u(t_0, \omega) = u_0(\omega), \qquad (2.5.1)$$

existing on $[t_0, t_0 + a)$;

(H$_4$) $m \in C[[t_0, t_0 + a), R[\Omega, R^n]]$, $(t, m(t)) \in E$ w.p. 1,

$$m(t_0, \omega) \leq u_0(\omega) \quad \text{w.p. 1} \qquad (2.5.2)$$

and

$$D^+ m(t, \omega) \leq g(t, m(t, \omega), \omega) \qquad \text{a.e. in} \quad t \in [t_0, t_0 + a). \qquad (2.5.3)$$

Then

$$m(t, \omega) \leq r(t, \omega) \qquad \text{for} \quad t \in [t_0, t_0 + a). \qquad (2.5.4)$$

*Proof.* Let $u(t, \omega, \varepsilon)$ be any sample solution of

$$u'(t, \omega, \varepsilon) = g(t, u(t, \omega, \varepsilon), \omega) + \varepsilon, \qquad u(t_0, \omega) = u_0(\omega) + \varepsilon$$

for $\varepsilon > 0$ sufficiently small. Then by Theorem 2.3.1, we have

$$m(t, \omega) < u(t, \omega, \varepsilon), \qquad t \in [t_0, t_0 + a).$$

Since $\lim_{\varepsilon \to 0} u(t, \omega, \varepsilon) \equiv r(t, \omega)$ on every closed interval contained in $[t_0, t_0 + a)$, the proof is complete.

**Corollary 2.5.1.** If in Theorem 2.5.1 inequalities (2.5.2) and (2.5.3) are reversed, then conclusion (2.5.4) is replaced by

$$m(t, \omega) \geq \rho(t, \omega), \qquad t \in [t_0, t_0 + a),$$

where $\rho(t, \omega)$ is the sample minimal solution of (2.5.1).

## 2.6. UNIQUENESS AND CONTINUOUS DEPENDENCE

We shall first give a simple result that assures the uniqueness of the solutions of (2.2.1).

**Theorem 2.6.1.** Assume that

(i) $f \in M[J \times \bar{B}(z, \rho), R[\Omega, R^n]]$ and $f(t, x, \omega)$ is sample continuous in $x$ for each $t \in J$;

(ii) $g \in M[J \times [0, 2\alpha], R[\Omega, R_+]]$, $g(t, u, \omega)$ is sample continuous in $u$ for each $t \in J$, and $g(t, u(t), \omega)$ is sample Lebesgue-integrable whenever $u(t)$ is sample absolutely countinuous, where $\alpha = \rho$;

(iii) $u(t) \equiv 0$ is the unique solution w.p. 1 of the scalar random differential equation

$$u'(t, \omega) = g(t, u(t, \omega), \omega), \qquad u(t_0, \omega) = 0, \qquad (2.6.1)$$

existing on $[t_0, t_0 + b] \subset J$;

(iv) for $(t, x), (t, y) \in J \times \bar{B}(z, \rho)$,

$$\|f(t, x, \omega) - f(t, x, \omega)\| \leq g(t, \|x - y\|, \omega); \qquad (2.6.2)$$

(v) $x_0 \in \bar{B}(z, \tfrac{1}{2}\rho)$;

(vi) $K \in IB[J, R[\Omega, R_+]]$ satisfies

$$\|f(t, x, \omega)\| \leq K(t, \omega) \qquad \text{for} \quad (t, x) \in J \times \bar{B}(z, \rho).$$

Then the differential system (2.2.1) has a unique solution on $[t_0, t_0 + b]$.

*Proof.* From (i), (v), (vi) and by using Theorem 2.2.1, it is obvious that (2.2.1) has a solution $x(t, \omega) = x(t, t_0, x_0, \omega)$. Let $x(t, \omega) = x(t, t_0, x_0, \omega)$ and $y(t, \omega) = y(t, t_0, x_0, \omega)$ be two solutions of (2.2.1) on $[t_0, t_0 + b]$. Define $m(t, \omega) = \|x(t, \omega) - y(t, \omega)\|$. Clearly, $m(t, \omega)$ is sample absolutely continuous on $[t_0, t_0 + b]$ since $x(t, \omega)$ is sample absolutely continuous and $\|x\|$ is Lipschitzian in $x$. Then

$$
\begin{aligned}
m'(t, \omega) &\leq \|x'(t, \omega) - y'(t, \omega)\| \\
&= \|f(t, x(t, \omega), \omega) - f(t, y(t, \omega), \omega)\| \\
&\leq g(t, m(t, \omega), \omega),
\end{aligned}
$$

using (2.6.2). Note that $m(t_0, \omega) = 0$ w.p. 1. By Theorem 2.5.1, we have

$$
m(t, \omega) \leq r(t, \omega) \qquad \text{on} \quad [t_0, t_0 + b],
$$

where $r(t, \omega)$ is the maximal solution process of (2.6.1). This, together with assumption (iii), implies that $m(t, \omega) \equiv 0$ on $[t_0, t_0 + b]$ w.p. 1. Thus the proof of the theorem is complete.

**Corollary 2.6.1.**   The function $g(t, u, \omega) = K(t, \omega)u$ is admissible in Theorem 2.6.1. In this case, condition (2.6.2) reduces to the well-known local Lipschitz condition.

**Remark 2.6.1.**   Under the conditions of Theorem 2.6.1 with $g(t, u, \omega) = K(t, \omega)u$, the proof of the existence theorem (Theorem 2.2.1) can be simplified by following the method of successive approximation. That is, define

$$
x_0(t, \omega) = x_0(\omega) \qquad \text{on} \quad J
$$

and

$$
x_{m+1}(t, \omega) = x_0(\omega) + \int_{t_0}^{t} f(s, x_m(s, \omega), \omega)\, ds, \qquad m = 1, 2, \ldots . \quad (2.6.3)
$$

In this case, the sequence $x_m(t)$ itself converges to the solution process uniquely. So the argument used to get the convergent subsequence in the proof of Theorem 2.2.1 is not necessary. Furthermore, assumption (vi) is not required.

**Remark 2.6.2.**   In addition to the hypotheses of Theorem 2.2.1, if we assume that the initial-value problem (2.2.1) has a unique solution if it exists, then the argument used to select the convergent subsequence can be modified. Further, note that the uniqueness assumption on (2.2.1) seems to be reasonable because the system (2.2.1) that describes the dynamic model of a practical system is well-defined.

In formulating a mathematical model for a physical or biological system, we make errors in constructing the functions $f(t, x, \omega)$ as well as errors in initial conditions. For mathematical purposes, it is sufficient to know that the change in the solutions can be made arbitrarily small by making arbitrary small changes in the differential equations and the initial values.

The following theorem establishes the continuous dependence of solutions on the parameters.

**Theorem 2.6.2.**  Assume that

$(H_1)_\lambda$   $f \in M[J \times D \times \Lambda, R[\Omega, R^n]]$, $f(t, x, \lambda, \omega)$ is sample continuous in $(x, \lambda)$ for each fixed $t$, $D$ is an open set in $R^n$, and $\Lambda$ is an open parameter $\lambda$-set in $R^m$;

$(H_2)_\lambda$   $g \in M[J \times R_+, R[\Omega, R_+]]$, $g(t, u, \omega)$ is sample continuous in $u$ for each $t \in J$, and $g(t, u(t), \omega)$ is sample Lebesgue-integrable whenever $u(t)$ is sample absolutely continuous;

$(H_3)_\lambda$   $u(t) \equiv 0$ is the unique solution w.p. 1 of (2.6.1) such that $u(t_0) = 0$ w.p. 1, and the solutions $u(t, t_0, u_0, \omega)$ of (2.6.1) are sample continuous with respect $u_0$;

$(H_4)_\lambda$   for $(t, x, \lambda)$, $(t, y, \lambda) \in J \times D \times \Lambda$,

$$\|f(t, x, \lambda, \omega) - f(t, y, \lambda, \omega)\| \le g(t, \|x - y\|, \omega); \qquad (2.6.4)$$

$(H_5)_\lambda$   $K \in IB[J, R[\Omega, R_+]]$ and satisfies

$$\|f(t, x, \lambda, \omega)\| \le K(t, \omega) \qquad \text{for} \quad (t, x, \lambda) \in J \times \bar{E}_0 \times \Lambda$$

$$(2.6.5)$$

w.p. 1, where $\bar{E}_0$ is any compact subset of $D$.

Then, given any scalar number $\varepsilon > 0$ and $\lambda_0 \in \Lambda$, there exists a $\delta = \delta(\varepsilon) > 0$ such that for every $\lambda$, $\|\lambda - \lambda_0\| \le \delta(\varepsilon)$, and the differential equation

$$x'(t, \omega) = f(t, x(t, \omega), \lambda, \omega), \qquad x(t_0, \lambda, \omega) = x_0(\lambda, \omega), \qquad (2.6.6)$$

admits a unique solution $x(t, \lambda, \omega) = x(t, t_0, x_0(\lambda, \omega), \omega)$ satisfying

$$\|x(t, \lambda, \omega) - x(t, \lambda_0, \omega)\| < \varepsilon, \qquad t \in J. \qquad (2.6.7)$$

*Proof.*   Under hypotheses $(H_1)_\lambda$ and $(H_5)_\lambda$, because of Corollary 2.2.1, the initial-value problem (2.6.6) has at least one solution on $[t_0, t_0 + b]$ whenever $(t_0, x_0, \lambda)$ belongs to a compact subset of $J \times D \times \Lambda$. Furthermore, by $(H_2)_\lambda$–$(H_4)_\lambda$ and Theorem 2.6.1, $x(t, \lambda)$ is the unique solution process of (2.6.6). Let $\lambda_0$ be any element of $\Lambda$. Since the family $\{x(t, \lambda)\}$ is sample equicontinuous and equibounded, it follows that for almost all fixed $\omega \in \Omega$, there exists a sequence of $\lambda_n$ such that $\lambda_n \to \lambda_0$ as $n \to \infty$; depending on $\omega$, the uniform limit

$$\lim_{\lambda_n \to \lambda_0} x(t, \lambda_n, \omega) = x(t, \lambda_0, \omega) \qquad (2.6.8)$$

exists.

From the sample continuity of $f(t, x, \lambda, \omega)$ in $(x, \lambda)$ for fixed $t \in [t_0, t_0 + b]$, we get

$$f(t, x(t, \lambda_n, \omega), \lambda_n, \omega) \to f(t, x(t, \lambda_0, \omega), \lambda_0, \omega) \qquad (2.6.9)$$

as $n(\omega) \to \infty$ for fixed $\omega \in \Omega$ in (2.6.8). By the application of the Lebesgue dominated convergence theorem, we see that

$$\int_{t_0}^{t} f(s, x(s, \lambda_n, \omega), \lambda_n, \omega) \, ds \to \int_{t_0}^{t} f(s, x(s, \lambda_0, \omega), \lambda_0, \omega) \, ds. \quad (2.6.10)$$

It is obvious that for fixed $\omega \in \Omega$ in (2.6.8), $x(t, \lambda_0, \omega)$ is the solution of the deterministic system

$$x'(t, \lambda_0, \omega) = f(t, x(t, \lambda_0, \omega), \lambda_0, \omega), \qquad x(t_0, \lambda_0) = x_0(\lambda_0, \omega). \tag{2.6.11}$$

For $\lambda = \lambda_0$, the solution process of (2.6.6) is unique. Therefore, for every sequence $\lambda_n$ such that $\lambda_n \to \lambda_0$ as $n \to \infty$, the uniform limit

$$\lim_{\lambda_n \to \lambda_0} x(t, \lambda_n, \omega) = x(t, \lambda_0, \omega) \qquad \text{for fixed} \quad \omega \in \Omega \text{ in (2.6.8)} \quad (2.6.12)$$

exists, which implies that $\lim_{\lambda_n \to \lambda_0} x(t, \lambda_n, \omega) = x(t, \lambda_0, \omega)$ w.p. 1. Hence $x(t, \lambda_0, \omega)$ is a random process satisfying (2.6.11), which shows that the limits (2.6.8), (2.6.9), (2.6.10), and (2.6.12) hold a.s. Moreover, (2.6.12) and the selection of sequence $\lambda_n$ imply that

$$\lim_{\lambda \to \lambda_0} x(t, \lambda, \omega) = x(t, \lambda_0, \omega) \qquad \text{uniformly in} \quad t \in [t_0, t_0 + b] \quad \text{w.p. 1,}$$

which shows that for any given scalar number $\varepsilon > 0$ and $\lambda_0 \in \Lambda$, there exists a $\delta = \delta(\varepsilon) > 0$ such that for every $\lambda$, $\|\lambda - \lambda_0\| \le \delta(\varepsilon)$, the solution process $x(t, \lambda)$ of (2.6.6) satisfies (2.6.7) for $t \in [t_0, t_0 + b]$. Now, by using Theorem 2.2.2 and the argument used in the proof of that theorem, the solution can be extended on $J$. This completes the proof of Theorem 2.6.2.

**Remark 2.6.3.** Assume that all the hypotheses of Theorem 2.6.2 hold, except that hypotheses $(H_2)_\lambda$–$(H_4)_\lambda$ are replaced by the assumption that the initial-value problem (2.6.6) has a unique solution; then the conclusion of Theorem 2.6.2 remains valid.

**Remark 2.6.4.** We note that the parameter $\lambda$ in Theorem 2.6.2 can be replaced by the random parameter $\lambda \in R(\Omega, R^m]$. The proof remains the same with slight modifications.

Now we present a theorem which establishes the continuous dependence of the solution processes of (2.2.1) with respect to the initial conditions $(t_0, x_0)$.

**Theorem 2.6.3.** Assume that the hypotheses of Theorem 2.6.1 hold. Furthermore, the solutions $u(t, \omega)$ of (2.6.1) through $(t_0, u_0)$ are sample continuous with respect to the initial conditions $(t_0, u_0)$. Then the solutions

$x(t, t_0, x_0)$ of 2.2.1) through $(t_0, x_0)$ are a.s. unique and sample continuous with respect to the initial conditions $(t_0, x_0)$.

*Proof.*   The proof of the existence and uniqueness of solutions $x(t)$ of (2.2.1) follows from the proof of Theorem 2.6.1. The proof of the continuous dependence on initial conditions can be formulated on the basis of the proof of Theorem 2.6.2. Consider the following system of random differential equations

$$x'(t, t_0, x_0(\omega), \omega) = f(t, x(t, t_0, x_0(\omega), \omega), \omega), \qquad x(t_0, \omega) = x_0(\omega), \quad (2.6.13)$$

where $(t_0, x_0)$ is considered a random parameter. In view of the hypotheses of the theorem and the fact that the function $f(t, x, \omega)$ is independent of the parameter, it is very easy to see that (2.6.13) satisfies the hypotheses of Theorem 2.6.2. However, the formal proof of the continuous dependence of solutions of (2.2.1) on the initial conditions can be formulated as follows:

Let $x_1(t, \omega) = x(t, t_1, x_1(\omega), \omega)$ and $x_2(t, \omega) = x(t, t_2, x_2(\omega), \omega)$ be solutions of (2.2.1) through $(t_1, x_1)$ and $(t_2, x_2)$, respectively. Without loss of generality, assume that $t_2 \le t_1$. Define

$$m(t, \omega) = \|x_1(t, \omega) - x_2(t, \omega)\|, \qquad m(t_1, \omega) = \|x_1 - x_2(t_1, \omega)\|.$$

$$(2.6.14)$$

From (2.6.2), we obtain

$$m'(t, \omega) \le g(t, m(t, \omega), \omega), \qquad t \ge t_1. \tag{2.6.15}$$

From (2.6.14), (2.6.15), and Theorem 2.5.1, we have

$$m(t, \omega) \le r(t, \omega), \qquad t \ge t_1, \tag{2.6.16}$$

where $r(t, \omega) = r(t, t_1, m(t_1, \omega), \omega)$ is the maximal solution process of (2.6.1) through $(t_1, m(t_1))$. Since $x_2(t, \omega)$ is sample continuous in $t$ and $r(t, t_0, u_0(\omega), \omega)$ is by hypothesis sample continuous in $(t_0, u_0)$, it follows from (iii), (2.6.16), and the definition of $m(t, \omega)$ that

$$\lim_{\substack{t_1 \to t_2 \\ x_1 \to x_2}} m(t, \omega) \le r(t, t_2, 0) \equiv 0 \quad \text{w.p. 1.}$$

Hence

$$\lim_{\substack{t_1 \to t_2 \\ x_1 \to x_2}} \|x_1(t, \omega) - x_2(t, \omega)\| = 0 \quad \text{w.p. 1,}$$

and the proof is complete.

**Lemma 2.6.1.**   Assume that

$(H_1)$   $f \in M[J \times D, R[\Omega, R^n]]$ and its sample derivative $\partial f / \partial x = (\partial / \partial x) f(t, x, \omega)$ exists and is sample continuous in $x$ for each $t \in J$, where $D$ is an open convex set in $R^n$. Then

$$f(t, y, \omega) - f(t, x, \omega) = \int_0^1 \left[ \frac{\partial}{\partial x} f(t, sy + (1 - s)x, \omega) \right](y - x) \, ds.$$

*Proof.*   Setting

$$F(s, \omega, t) = f(t, sy + (1 - s)x, \omega), \qquad 0 \leq s \leq 1,$$

the convexity of $D$ implies that $F(s, \omega, t)$ is well-defined. It is obvious that $F(s, \omega, t)$ is sample continuous in $s$. Since $f(t, x, \omega)$ is sample differentiable in $x$, it follows that $F(s, \omega, t)$ is sample differentiable in $s \in [0, 1]$. Moreover, its sample derivative $\partial F / \partial s$ is a sample continuous random variable. Hence

$$\frac{\partial}{\partial s} F(s, \omega, t) = \frac{\partial}{\partial x} f(t, sy + (1 - s)x, \omega)(y - x). \qquad (2.6.17)$$

Since $F(1, \omega, t) = f(t, y, \omega)$ and $F(0, \omega, t) = f(t, x, \omega)$, the result follows by sample integrating (2.6.17) in the sense of Riemann with respect to $s$ from 0 to 1.

In the following we shall prove that the solution process $x(t, t_0, x_0, \omega)$ of (2.2.1) is sample differentiable with respect to the initial conditions $(t_0, x_0)$ and that the sample derivatives $(\partial / \partial x_0)x(t, t_0, x_0, \omega)$ and $(\partial / \partial t_0)x(t, t_0, x_0, \omega)$ exist and satisfy the equation of variation of (2.2.1) along the solution process $x(t, t_0, x_0, \omega)$.

**Theorem 2.6.4.**   Assume that

(H$_1$)   $f \in M[J \times R^n, R[\Omega, R^n]]$ and its sample derivative

$$\frac{\partial f}{\partial x} = \frac{\partial}{\partial x} f(t, x, \omega)$$

exists and is sample continuous in $x$ for each $t \in J$;

(H$_2$)   $K \in IB[J, R[\Omega, R_+]]$ and satisfies

$$\|f_x(t, x, \omega)\| \leq K(t, \omega) \qquad \text{for} \quad (t, x) \in J \times \bar{B}(z, \rho), \quad (2.6.18)$$

where $f_x(t, x, \omega) = (\partial / \partial x)f(t, x, \omega)$;

(H$_3$)   the solution $x(t, \omega) = x(t, t_0, x_0(\omega), \omega)$ of (2.2.1) exists for $t \geq t_0$.

Then

(a)   the sample derivative $(\partial / \partial x_{k_0})x(t, t_0, x_0(\omega), \omega)$ exists for all $k = 1, 2, 3, \ldots, n$ and satisfies the systems of linear random differential equations

$$y'(t, \omega) = f_x(t, x(t, t_0, x_0(\omega), \omega), \omega)y(t, \omega), \quad y(t_0, \omega) = e_k, \qquad (2.6.19)$$

where $e_k = (e_k^1, e_k^2, \ldots, e_k^k, \ldots, e_k^n)^T$ is an $n$-vector such that $e_k^j = 0$ if $j \neq k$ and $e_k^k = 1$ and $x_{k0}$ is the $k$th component of $x_0 = (x_{10}, x_{k0}, \ldots, x_{n0})^T$ and

(b)   the sample derivative $(\partial/\partial t_0)x(t, t_0, x_0(\omega), \omega)$ exists and satisfies (2.6.19) with

$$\frac{\partial}{\partial t_0}\, x(t, t_0, x_0(\omega), \omega) = -\Phi(t, t_0, x_0(\omega), \omega)\dot{f}(t_0, x_0(\omega), \omega) \qquad (2.6.20)$$

whenever $f(t_0, x_0(\omega), \omega)$ exists w.p. 1, where $\Phi(t, t_0, x_0, \omega)$ is the fundamental matrix solution process of (2.6.19). Moreover, $\Phi(t, t_0, x_0, \omega)$ satisfies the random matrix differential equation

$$X' = f_x(t, x(t), \omega)X, \qquad X(t_0, \omega) = \text{unit matrix.} \qquad (2.6.21)$$

*Proof.*   First we shall show that (a) holds. From hypotheses $(H_1)$ and $(H_2)$, Lemma 2.6.1, and Corollary 2.6.1, it is obvious that the solution $x(t, \omega)$ in $(H_3)$ is unique. Moreover, by Theorem 2.6.3, the solutions $x(t, t_0, x_0)$ are continuous with respect to the initial conditions $(t_0, x_0)$. For small $\lambda$, $x(t, \lambda, \omega) = x(t, t_0, x_0(\omega) + \lambda e_k, \omega)$ and $x(t, \omega) = x(t, t_0, x_0(\omega), \omega)$ are solution processes of (2.2.1) through $(t_0, x_0 + \lambda e_k)$ and $(t_0, x_0)$, respectively. From the continuous dependence on initial conditions, it is clear that

$$\lim_{\lambda \to 0} x(t, \lambda, \omega) = x(t, t_0, \omega) \qquad \text{uniformly on}\quad J\quad \text{w.p. 1.} \qquad (2.6.22)$$

Set

$$\Delta x_\lambda(t, \omega) = \frac{[x(t, \lambda, \omega) - x(t, \omega)]}{\lambda}, \qquad \Delta x_\lambda(t_0, \omega) = e_k, \qquad \lambda \neq 0, \qquad (2.6.23)$$

$x = x(t, \omega)$ and $y = x(t, \lambda, \omega)$. This, together with the application of Lemma 2.6.1, yields

$$\Delta x'_\lambda(t, \omega) = \int_0^1 f_x(t, sy + (1 - s)x, \omega)\,\Delta x_\lambda(t, \omega)\,ds. \qquad (2.6.24)$$

Define

$$f_x(t, x(t, \omega), \lambda, \omega) = \int_0^1 f_x(t, sx(t, \lambda, \omega) + (1 - s)x(t, \omega), \omega)\,ds. \qquad (2.6.25)$$

Note that the integral is the sample Riemann integral. In view of $(H_1)$, $f_x(t, x, \lambda, \omega)$ is a product-measurable random process which is sample continuous in $(x, \lambda)$ for fixed $t$. Furthermore,

$$\|f_x(t, x, \lambda, \omega)\| \leq K(t, \omega) \qquad \text{for}\quad (t, x, \lambda) \in J \times \bar{B}(z, \rho) \times \Lambda,$$

where $\Lambda \subset R$ is an open neighborhood of $0 \in R$. From (2.6.22) and the sample continuity of $f_x(t, x, \omega)$, we have

$$\lim_{\lambda \to 0} f_x(t, sx(t, \lambda, \omega) + (1 - s)x(t, \omega), \omega) = f_x(t, x(t, \omega), \omega) \qquad (2.6.26)$$

uniformly in $s \in [0, 1]$ w.p. 1. This, together with (2.6.25), yields

$$\lim_{\lambda \to 0} f_x(t, x(t, \omega), \lambda, \omega) = f_x(t, x(t, \omega), \omega) \quad \text{w.p. 1.} \qquad (2.6.27)$$

From (2.6.23) and (2.6.25), relation (2.6.24) reduces to

$$\Delta x'_\lambda(t, \omega) = f_x(t, x(t, \omega), \lambda, \omega) \Delta x_\lambda(t, \omega), \qquad \Delta x_\lambda(t_0, \omega) = e_k. \quad (2.6.28)$$

It is obvious that the initial-value problem (2.6.28) satisfies hypotheses $(H_1)_\lambda$–$(H_5)_\lambda$, and hence by the application of Theorem 2.6.2, we have

$$\lim_{\lambda \to 0} \Delta x_\lambda(t, \omega) = y(t, \omega) \quad \text{uniformly on} \quad J \quad \text{w.p. 1,} \qquad (2.6.29)$$

where $y(t, \omega)$ is the solution process of (2.6.19). Because of (2.6.23), we note that the limit of $\Delta x_\lambda(t, \omega)$ in (2.6.29) is equivalent to the derivative $(\partial/\partial x_{k0})x(t, t_0, x_0(\omega), \omega)$. Hence $(\partial/\partial x_{k0})x(t, t_0, x_0(\omega), \omega)$ exists and is the solution process of (2.6.19). This is true for every $k = 1, 2, \ldots, n$. Thus $(\partial/\partial x_0)x(t, t_0, x_0(\omega), \omega)$ is the fundamental random matrix solution process of (2.6.19) and satisfies the random matrix equation (2.6.21). $(\partial/\partial x_0)x(t, t_0, x_0(\omega), \omega)$ is denoted by $\Phi(t, t_0, x_0(\omega), \omega)$. This completes the proof of (a) and the last part of (b).

To prove the first part of (b), define

$$\Delta \hat{x}_\lambda(t, \omega) = \frac{x(t, \lambda, \omega) - x(t, \omega)}{\lambda}, \qquad \Delta \hat{x}_\lambda(t_0, \omega) = \frac{x(t_0, \lambda, \omega) - x_0(\omega)}{\lambda}, \quad (2.6.30)$$

where $x(t, \lambda, \omega) = x(t, t_0 + \lambda, x_0(\omega), \omega)$ and $x(t, \omega) = x(t, t_0, x_0(\omega), \omega)$ are solution processes of (2.2.1) through $(t_0 + \lambda, x_0)$ and $(t_0, x_0)$, respectively. Again, by imitating the proof of part (a) and by replacing $\Delta x_\lambda(t, \omega)$ by $\Delta \hat{x}_\lambda(t, \omega)$, one can conclude that the sample derivative $(\partial/\partial t_0)x(t, t_0, x_0(\omega), \omega)$ exists and is the solution of (2.6.19) whenever

$$\lim_{\lambda \to 0} \Delta \hat{x}_\lambda(t_0, \omega) \text{ exists and is equal to } -f(t_0, x_0(\omega), \omega). \qquad (2.6.31)$$

We shall show that (2.6.31) is true. By uniqueness of solutions, we have

$$\Delta \hat{x}_\lambda(t_0, \omega) = (x(t_0, t_0 + \lambda, x_0(\omega), \omega) - x(t_0 + \lambda, t_0 + \lambda, x_0(\omega), \omega))/\lambda$$

$$= -\frac{x(t_0 + \lambda, t_0 + \lambda, x_0(\omega), \omega) - x(t_0, t_0 + \lambda, x_0(\omega), \omega)}{\lambda}$$

$$= -\frac{1}{\lambda} \int_{t_0}^{t_0 + \lambda} f(s, x(s, t_0 + \lambda, x_0(\omega), \omega), \omega) \, ds.$$

This, together with the sample Lebesgue integrability of

$$f(s, x(s, t_0 + \lambda, x_0(\omega), \omega), \omega), \qquad \text{implies that} \quad \lim_{\lambda \to 0} \Delta \hat{x}_\lambda(t_0, \omega)$$

exists and is equal to $-f(t_0, x_0(\omega), \omega)$ w.p. 1. Now by following the proof of the deterministic Theorem 2.5.3 in Lakshmikantham and Leela [59], we can conclude that

$$\frac{\partial}{\partial t_0} x(t, t_0, x_0(\omega), \omega) = -\Phi(t, t_0, x_0(\omega), \omega) f(t_0, x_0(\omega), \omega) \quad \text{w.p. 1.}$$

This completes the proof of the first part of (b). Hence the proof of the theorem is complete.

**Remark 2.6.5.** It is easy to see that the fundamental random matrix $\Phi(t, t_0, x_0, \omega)$ is sample continuous in $(t_0, x_0)$ for fixed $t$. This is because of the application of Theorem 2.6.3 and the facts that $f_x(t, x, \omega)$ is sample continuous in $x$ for fixed $t$ and the solution process $x(t, t_0, x_0, \omega)$ is sample continuous in $(t_0, x_0)$ for fixed $t$.

### 2.7. THE METHOD OF VARIATION OF PARAMETERS

Let us present some elementary results about the linear random differential system

$$x'(t, \omega) = A(t, \omega)x(t, \omega), \qquad x(t_0, \omega) = y_0(\omega), \tag{2.7.1}$$

where $A(t, \omega) = (a_{ij}(t, \omega))$ is a product-measurable random matrix function defined on $J \times \Omega$ into $R^{n^2}$. Assume also that $A(t, \omega)$ is a.s. sample Lebesgue-integrable on $J$. Let $\Phi(t, \omega)$ be the $n \times n$ matrix whose columns are the $n$-vector solutions of (2.7.1) with the initial condition $y_0(\omega) = e_k$, $k = 1$, $2, \ldots, n$. Clearly, $\Phi(t_0, \omega) = $ unit matrix, and $\Phi(t, \omega)$ satisfies the random matrix differential equation

$$\Phi'(t, \omega) = A(t, \omega)\Phi(t, \omega), \qquad \Phi(t_0, \omega) = \text{unit matrix}. \tag{2.7.2}$$

The following result is analogous to the corresponding deterministic result, whose proof can be formulated similarly.

**Lemma 2.7.1.** Let $A(t, \omega)$ be a product-measurable random matrix function defined on $J \times (\Omega, \mathscr{F}, P)$ into $R^n$ and sample Lebesgue-integrable on $J$. Then the fundamental matrix solution process $\Phi(t, \omega)$ of (2.7.2) is a.s. nonsingular on $J$. Moreover,

$$\det \Phi(t, \omega) = \exp\left[ \int_{t_0}^{t} \text{tr } A(s, \omega) \, ds \right], \qquad t \in J, \tag{2.7.3}$$

where $\text{tr } A(t, \omega) = \sum_{i=1}^{n} a_{ii}(t, \omega)$ and det stands for a determinant of a matrix.

Consider the random perturbed system of the linear system (2.7.1),

$$y'(t,\omega) = A(t,\omega)y(t,\omega) + F(t, y(t,\omega), \omega), \qquad y(t_0,\omega) = y_0(\omega), \quad (2.7.4)$$

where $A(t,\omega)$ is as defined above and $F \in M[J \times R^n, R[\Omega, R^n]]$.

The following result establishes the integral representation for the solution processes of (2.7.4) in terms of the solution process of (2.7.1).

**Theorem 2.7.1.**   Assume that

(i)   $A(t,\omega)$ satisfies the hypotheses of Lemma 2.7.1,

(ii)   $F \in M[J \times R^n, R[\Omega, R^n]]$ and $F(t, x, \omega)$ is sample continuous in $x$ for fixed $t$,

(iii)   $K \in IB[J, R[\Omega, R_+]]$ and satisfies

$$\|F(t, x, \omega)\| \le K(t, \omega) \qquad \text{for} \quad (t, x) \in J \times \bar{B}(z, \rho).$$

Then any solution process $y(t,\omega) = y(t, t_0, y_0(\omega), \omega)$ of (2.7.4) satisfies the random integral equation

$$y(t,\omega) = x(t,\omega) + \int_{t_0}^{t} \Phi(t,\omega)\Phi^{-1}(s,\omega)F(s, y(s,\omega), \omega)\,ds, \qquad t \ge t_0, \quad (2.7.5)$$

where $\Phi(t,\omega)$ is the fundamental matrix solution of (2.7.1).

*Proof.*   Let $x(t,\omega) = \Phi(t,\omega)y_0(\omega)$ be a solution process of (2.7.1), existing for $t \ge t_0$. The method of variation of parameters requires the determination of an a.s. sample differentiable function $z(t,\omega)$ so that

$$y(t,\omega) = \Phi(t,\omega)z(t,\omega), \qquad z(t_0,\omega) = y_0(\omega), \quad\quad\quad (2.7.6)$$

is a solution process of (2.7.4). By sample differentiation with respect to $t$, we get

$$A(t,\omega)y(t,\omega) + F(t, y(t,\omega), \omega) = \Phi'(t,\omega)z(t,\omega) + \Phi(t,\omega)z'(t,\omega).$$

This, because of (2.7.2) and (2.7.6), yields

$$z'(t,\omega) = \Phi^{-1}(t,\omega)F(t, \Phi(t,\omega)z(y,\omega), \omega), \qquad z(t_0,\omega) = y_0(\omega). \quad (2.7.7)$$

From (i)–(iii), (2.7.3), and the fact that

$$\Phi^{-1}(t,\omega) = \left(\frac{(-1)^{i+j}\det(\Phi_{ij}(t,\omega))}{\det \Phi(t,\omega)}\right)^{\mathrm{T}},$$

the random function $\Phi^{-1}(t,\omega)F(t, y(t,\omega), \omega)$ satisfies the hypotheses corresponding to (ii) and (iii), where T stands for the transpose of a matrix and $\Phi_{ij}(t,\omega)$ is the matrix obtained from $\Phi(t,\omega)$ by deleting the $i$th and the $j$th columns. Therefore, the initial-value problem (2.7.7) has a solution process

$z(t, \omega)$ existing for $t \geq t_0$. Hence

$$z(t, \omega) = y_0(\omega) + \int_{t_0}^t \Phi^{-1}(s, \omega)F(s, \Phi(s, \omega)z(s, \omega), \omega)\,ds. \qquad (2.7.8)$$

From (2.7.6) and (2.7.8), we obtain

$$y(t, \omega) = \Phi(t, \omega)y_0(\omega) + \int_{t_0}^t \Phi(t, \omega)\Phi^{-1}(s, \omega)F(s, y(s, \omega), \omega)\,ds.$$

This completes the proof of the theorem.

In the following, we shall present a result similar to Theorem 2.7.1 with respect to

$$y'(t, \omega) = f(t, y(t, \omega), \omega) + F(t, y(t, \omega), \omega), \qquad y(t_0, \omega) = x_0(\omega), \quad (2.7.9)$$

which is a perturbation of a nonlinear system (2.2.1).

**Theorem 2.7.2.** Assume that

(i)  $f, F \in M[J \times R^n, R[\Omega, R^n]]$ and $f(t, x, \omega)$ and $F(t, x, \omega)$ are a.s. sample continuous in $x$ for fixed $t \in J$;

(ii)  $K \in IB[J, R[\Omega, R_+]]$ and satisfies

$$\|f(t, x, \omega)\| + \|F(t, x, \omega)\| \leq K(t, \omega) \qquad \text{for} \quad (t, x) \in J \times \bar{B}(z, \rho);$$

(iii)  the initial-value problem (2.2.1) has a unique solution process $x(t, \omega)$, existing for $t \geq t_0$;

(iv)  the sample derivative $\Phi(t, t_0, x_0, \omega) = (\partial/\partial x_0)x(t, t_0, x_0(\omega), \omega)$ exists and $\Phi(t, t_0, x_0, \omega)$ is sample continuous in $x_0$ for fixed $(t, t_0)$ and is product-measurable in $(t, \omega)$ for fixed $t_0$;

(v)  the inverse $\Phi^{-1}(t, t_0, x_0, \omega)$ of random matrix function $\Phi(t, t_0, x_0, \omega)$ exists and is product-measurable in $(t, \omega)$ and sample continuous in $x_0$ for fixed $(t, t_0)$;

(vi)  $K_1 \in IB[J, R[\Omega, R_+]]$ and satisfies

$$\|\Phi^{-1}(t, t_0, x_0, \omega)\| \leq K_1(t, \omega) \qquad \text{for} \quad (t, x) \in J \times \bar{B}(z, \rho).$$

Then any solution process $y(t, \omega) = y(t, t_0, x_0(\omega), \omega)$ of (2.7.9) satisfies the relation

$$y(t, t_0, x_0(\omega), \omega) = x\Bigg(t, t_0, x_0(\omega)$$

$$+ \int_{t_0}^t \Phi^{-1}(s, t_0, z(s, \omega), \omega)F(s, y(s, \omega), \omega)\,ds, \omega\Bigg), \quad (2.7.10)$$

as far as $z(t, \omega)$ exists to the right of $t_0$, w.p. 1, where $z(t, \omega)$ is a solution process of the initial-value problem

$$z'(t, \omega) = \Phi^{-1}(t, t_0, z(t,\omega), \omega)F(t, x(t, t_0, z(t,\omega), \omega), \omega),$$
$$z(t_0, \omega) = x_0(\omega). \tag{2.7.11}$$

*Proof.*   In view of hypotheses (i), (ii), (iv), (v), and (vi) and Theorem 2.2.2, the initial-value problems (2.2.1) and (2.7.11) have solution processes $x(t, t_0, x_0, \omega)$ and $z(t, t_0, x_0, \omega)$, respectively. The method of variation of parameters requires the determination of a random function $z(t, \omega)$ so that

$$y(t, t_0, x_0(\omega), \omega) = x(t, t_0, z(t,\omega), \omega), \qquad z(t_0, \omega) = x_0(\omega), \quad (2.7.12)$$

is a solution process of (2.7.9). The sample differentiation yields

$$f(t, y(t, \omega), \omega) + F(t, y(t, \omega), \omega) = x'(t, t_0, z(t, \omega), \omega)$$

$$+ \frac{\partial}{\partial x_0} x(t, t_0, z(t, \omega), \omega)z'(t, \omega).$$

This, together with (iv), (v), and (2.7.12), gives

$$z'(t, \omega) = \Phi^{-1}(t, t_0, z(t, \omega), \omega)F(t, x(t, t_0, z(t, \omega), \omega), \omega),$$
$$z(t_0, \omega) = x_0(\omega),$$

which implies

$$z(t, \omega) = x_0(\omega) + \int_{t_0}^t \Phi^{-1}(s, t_0, z(s, \omega), \omega)F(s, x(s, t_0, z(s, \omega), \omega), \omega)\, ds. \tag{2.7.13}$$

By (2.7.12) and (2.7.13), we have

$$y(t, t_0, x_0(\omega), \omega) = x\left(t, t_0, x_0(\omega)\right.$$

$$\left. + \int_{t_0}^t \Phi^{-1}(s, t_0, z(s, \omega), \omega)F(s, y(s, \omega), \omega)\, ds, \omega\right),$$

which proves the theorem.

**Corollary 2.7.1.**   Under the hypotheses of Theorem 2.7.2, the following relation is also valid:

$$y(t, t_0, x_0(\omega), \omega) = x(t, t_0, x_0(\omega), \omega)$$

$$+ \int_{t_0}^t \Phi(t, t_0, z(s, \omega), \omega)\Phi^{-1}(s, t_0, z(s, \omega), \omega)$$

$$\times F(s, y(s, t_0, x_0(\omega), \omega), \omega)\, ds, \tag{2.7.14}$$

where $z(t, \omega)$ is any solution process of (2.7.11).

*Proof.*    For $t_0 \leq s \leq t$, we have

$$\frac{d}{ds} x(t, t_0, z(s, \omega), \omega) = \frac{\partial}{\partial x_0} x(t, t_0, z(s, \omega), \omega) z'(s, \omega)$$

$$= \Phi(t, t_0, z(s, \omega), \omega) \Phi^{-1}(s, t_0, z(s, \omega), \omega)$$
$$\times F(s, x(s, t_0, z(s, \omega), \omega), \omega),$$

which implies

$$x(t, t_0, z(t, \omega), \omega) = x(t, t_0, x_0(\omega), \omega)$$
$$+ \int_{t_0}^{t} \Phi(t, t_0, z(s, \omega), \omega) \Phi^{-1}(s, t_0, z(s, \omega), \omega)$$
$$\times F(s, x(s, t_0, z(s, \omega), \omega), \omega), \omega)\, ds.$$

This, together with (2.7.12), yields

$$y(t, t_0, x_0(\omega), \omega) = x(t, t_0, x_0(\omega), \omega)$$
$$+ \int_{t_0}^{t} \Phi(t, t_0, z(s, \omega), \omega) \Phi^{-1}(s, t_0, z(s, \omega), \omega)$$
$$\times F(s, y(s, t_0, x_0(\omega), \omega), \omega), \omega)\, ds,$$

establishing relation (2.7.14).

**Remark 2.7.1.**    Relations (2.7.10) and (2.7.14) provide two different forms of the variation of parameters formula for the solution process of (2.7.9).

In the following, we give a result which is an analogue of a known deterministic nonlinear variation of constants type result due to Alekseev. See Lakshmikantham and Leela [59]. Furthermore, we shall show that the obtained nonlinear variation of constants formula is equivalent to formula (2.7.14).

**Theorem 2.7.3.**    Assume that the random functions $f$ and $F$ in (2.7.9) satisfy the hypotheses of Theorem 2.6.4 and Theorem 2.7.2, respectively. Then any solution process $y(t, \omega)$ of (2.7.9) satisfies formulas (2.7.10) and (2.7.14) and in addition the formula

$$y(t, t_0, x_0(\omega), \omega) = x(t, t_0, x_0(\omega), \omega)$$
$$+ \int_{t_0}^{t} \Phi(t, s, y(s, \omega), \omega) F(s, y(s, \omega), \omega)\, ds. \tag{2.7.15}$$

Moreover, (2.7.14) and (2.7.15) are equivalent if

$$\Phi(t, s, z(s, \omega), \omega) = \Phi(t, t_0, z(s, \omega), \omega) \Phi^{-1}(s, t_0, z(s, \omega), \omega).$$

*Proof.*    By Theorem 2.6.4 and Remark 2.6.4, the initial-value problem (2.2.1) has a unique solution process $x(t, \omega) = x(t, t_0, x_0(\omega), \omega)$. Furthermore,

$(\partial/\partial x_0)x(t, t_0, x_0(\omega), \omega) = \Phi(t, t_0, x_0(\omega), \omega)$ exists and is sample continuous in $(t, t_0, x_0)$, and $\Phi(t, t_0, x_0(\omega), \omega)$ is the fundamental random matrix solution process of (2.6.19), which implies that $\Phi^{-1}(t, t_0, x_0, \omega)$ exists and is sample continuous in $(t, t_0, x_0)$. Thus $f(t, x, \omega)$, $F(t, x, \omega)$, $\Phi(t, t_0, x_0, \omega)$, and $\Phi^{-1}(t, t_0, x_0, \omega)$ satisfy the hypotheses of Theorem 2.7.2. Therefore, the initial-value problem (2.7.9) satisfies formulas (2.7.10) and (2.7.14). To show that $y(t, \omega)$ satisfies (2.7.15), consider, for $t_0 \le s \le t$, $x(t, s, y(s, \omega), \omega)$ and $y(s, t_0, x_0(\omega), \omega)$ as solution processes of (2.2.1) and (2.7.9) through $(s, y(s, \omega))$ and $(t_0, x_0(\omega))$, respectively. Then

$$\frac{d}{ds}x(t, s, y(s, \omega), \omega) - \frac{\partial}{\partial x_0}x(t, s, y(s, \omega), \omega)y'(s, \omega)$$

$$+ \frac{\partial}{\partial t_0}x(t, s, y(s, \omega), \omega).$$

This, together with (2.6.20) and (2.7.9), yields

$$\frac{d}{ds}x(t, s, y(s, \omega), \omega) = \Phi(t, s, y(s, \omega), \omega)F(s, y(s, \omega), \omega), \qquad (2.7.16)$$

which implies

$$x(t, t, y(t, \omega), \omega) = x(t, t_0, x_0(\omega), \omega)$$

$$+ \int_{t_0}^{t} \Phi(t, s, y(s, \omega), \omega)F(s, y(s, \omega), \omega)\,ds.$$

By the uniqueness of the solution process of (2.2.1), we have

$$y(t, t_0, x_0(\omega), \omega) = x(t, t_0, x_0(\omega), \omega)$$

$$+ \int_{t_0}^{t} \Phi(t, s, y(s, \omega), \omega)F(s, y(s, \omega), \omega)\,ds$$

proving relation (2.7.15). We next show that formulas (2.7.14) and (2.7.15) are equivalent if and only if

$$\Phi(t, s, y(s, \omega), \omega) = \Phi(t, t_0, z(s, \omega), \omega)\Phi^{-1}(s, t_0, z(s, \omega), \omega).$$

For $t_0 \le s \le t$, we note that

$$x(t, s, y(s, \omega), \omega) = x(t, t_0, z(s, \omega), \omega), \qquad (2.7.17)$$

where $x(t, \omega)$ and $y(s, \omega)$ are as defined above and $z(s, \omega)$ is a solution process of (2.7.11) through $(t_0, x_0)$. Then by sample differentiation with respect to $s$, (2.7.17), and the uniqueness of the solution of (2.2.1), after substituting for $y'(s, \omega)$ and $z'(s, \omega)$, we get

$$\Phi(t, s, y(s, \omega), \omega)F(s, y(s, \omega), \omega)$$

$$= \Phi(t, t_0, z(s, \omega), \omega)\Phi^{-1}(s, t_0, z(s, \omega), \omega)F(s, y(s, \omega), \omega),$$

which is equivalent to

$$\Phi(t, s, y(s, \omega), \omega) = \dot\Phi(t, t_0, z(s, \omega), \omega)\Phi^{-1}(s, t_0, z(s, \omega), \omega).$$

This, together with (2.7.16), shows that (2.7.14) and (2.7.15) are equivalent if and only if

$$\Phi(t, s, y(s, \omega), \omega) = \Phi(t, t_0, z(s, \omega), \omega)\Phi^{-1}(s, t_0, z(s, \omega), \omega).$$

This completes the proof of the theorem.

The following theorem gives the relationship between the solution process of (2.2.1) and the fundamental random matrix solution process of (2.6.19).

**Theorem 2.7.4.** Assume that the hypotheses of Theorem 2.6.4 hold. Furthermore, assume that $x(t, t_0, x_0, \omega)$ and $x(t, t_0, y_0, \omega)$ are the solution processes of (2.2.1) through $(t_0, x_0)$ and $(t_0, y_0)$, respectively, existing for $t \geq t_0$, such that $x_0, y_0$ belong to a convex subset of $R^n$. Then for $t \geq t_0$,

$$x(t, t_0, y_0, \omega) - x(t, t_0, x_0, \omega)$$
$$= \int_0^1 \Phi(t, t_0, x_0 + s(y_0 - x_0), \omega)\,ds(y_0 - x_0). \qquad (2.7.18)$$

*Proof.* Since $x_0$, $y_0$ belong to a convex subset of $R^n$ w.p. 1, $x(t, t_0, x_0 + s(y_0 - x_0), \omega)$ is defined for $0 \leq s \leq 1$. Thus

$$\frac{d}{ds} x(t, t_0, x_0 + s(y_0 - x_0), \omega) = \Phi(t, t_0, x_0 + s(y_0 - x_0), \omega)(y_0 - x_0),$$

and hence integration from 0 to 1 yields (2.7.18).

## 2.8. RANDOM LYAPUNOV FUNCTIONS

Historically, the Lyapunov second method has played a very significant role in the qualitative and quantitative analysis of solutions of systems of differential equations. In the following, by employing the concept of random vector Lyapunov functions and the theory of random differential inequalities, we shall develop some results which furnish very general comparison theorems for random differential systems.

Let $B(\rho) = [x \in R^n: \|x\| < \rho]$, where $\rho$ is a positive number. Consider the system of random differential equations

$$x'(t, \omega) = f(t, x(t, \omega), \omega), \qquad x(t_0, \omega) = x_0(\omega), \qquad (2.8.1)$$

where $x \in R^n$, $f \in M[R_+ \times B(\rho), R[\Omega, R^n]]$, and $f$ is smooth enough to guarantee the existence of a sample solution process $x(t, \omega) = x(t, t_0, x_0, \omega)$ of (2.8.1) for $t \geq t_0$.

Assume that $V \in C[R_+ \times B(\rho), R[\Omega, R^m]]$, and define

$$D^+_{(2.8.1)}V(t, x, \omega) \equiv \limsup_{h \to 0^+}(1/h)[V(t + h, x + hf(t, x, \omega), \omega) - V(t, x, \omega)].$$

$$(2.8.2)$$

It is clear that $D^+_{(2.8.1)}V(t, x, \omega)$ exists for all $(t, x) \in R_+ \times B(\rho)$ and is a product-measurable random process.

We shall now formulate the following fundamental comparison theorem in the framework of random vector Lyapunov functions.

**Theorem 2.8.1.**  Suppose that

(i)  $g \in M[R_+ \times R^m, R[\Omega, R^m]]$ and $g(t, u, \omega)$ is sample continuous and quasi-monotone nondecreasing in $u$ for fixed $t \in R_+$;

(ii)  $r(t, \omega) = r(t, t_0, u_0, \omega)$ is the maximal solution process of the random differential system

$$u'(t, \omega) = g(t, u(t, \omega), \omega), \qquad u(t_0, \omega) = u_0(\omega), \qquad (2.8.3)$$

existing for $t \geq t_0$;

(iii)  $V \in C[R_+ \times R^n, R[\Omega, R^m]]$, $V(t, x, \omega)$ is locally Lipschitzian in $x$ w.p. 1, and for $(t, x) \in R_+ \times R^n$,

$$D^+_{(2.8.1)}V(t, x, \omega) \leq g(t, V(t, x, \omega), \omega); \qquad (2.8.4)$$

(iv)  $x(t, \omega) = x(t, t_0, x_0, \omega)$ is any sample solution process of (2.8.1) such that

$$V(t_0, x_0(\omega), \omega) \leq u_0(\omega), \qquad (2.8.5)$$

existing for $t \geq t_0$.

Then

$$V(t, x(t, \omega), \omega) \leq r(t, \omega) \qquad \text{for} \quad t \geq t_0. \qquad (2.8.6)$$

*Proof.*  Set

$$m(t, \omega) = V(t, x(t, \omega), \omega), \qquad m(t_0, \omega) = V(t_0, x_0(\omega), \omega). \qquad (2.8.7)$$

Since $x(t, \omega)$ is a sample solution process of (2.8.1) and $V \in C[R_+ \times R^n, R[\Omega, R^m]]$, it is obvious that $m(t, \omega)$ is sample continuous for $t \geq t_0$. For small $h > 0$, we have

$$m(t + h, \omega) - m(t, \omega)$$
$$= V(t + h, x(t + h, \omega), \omega) - V(t, x(t, \omega), \omega)$$
$$= V(t + h, x(t + h, \omega), \omega) - V(t + h, x(t, \omega) + hf(t, x(t, \omega), \omega), \omega)$$
$$\quad + V(t + h, x(t, \omega) + hf(t, x(t, \omega), \omega), \omega) - V(t, x(t, \omega), \omega)$$
$$\leq K\|x(t + h, \omega) - x(t, \omega) - hf(t, x(t, \omega), \omega)\|e$$
$$\quad + V(t + h, x(t, \omega) + hf(t, x(t, \omega), \omega), \omega) - V(t, x(t, \omega), \omega),$$

where $K$ is the local Lipschitz constant relative to $V(t, x, \omega)$ and $e = (1, 1, \ldots, 1)^{\mathrm{T}}$. This, together with (2.8.2), (2.8.4), (2.8.7), and the sample continuity of $m(t, \omega)$, yields the inequality

$$D^{+}m(t, \omega) \leq g(t, m(t, \omega), \omega), \qquad t \geq t_0. \qquad (2.8.8)$$

Hence by Theorem 2.5.1, we have

$$m(t, \omega) \leq r(t, \omega), \qquad t \geq t_0.$$

The proof is complete.

**Corollary 2.8.1.**   Assume that the hypotheses of Theorem 2.8.1 hold with $g(t, u, \omega) \equiv 0$ w.p. 1. Then the function $V(t, x(t, \omega), \omega)$ is sample nondecreasing in $t$ and

$$V(t, x(t, \omega), \omega) \leq V(t_0, x_0(\omega), \omega), \qquad t \geq t_0.$$

The following variant of Theorem 2.8.1 is often useful in applications.

**Theorem 2.8.2.**   Let the hypotheses of Theorem 2.8.1 hold except that inequality (2.8.4) is replaced by

$$A(t)D^{+}_{(2.8.1)}V(t, x, \omega) + A'(t)V(t, x, \omega) \leq g(t, A(t)V(t, x, \omega), \omega) \qquad (2.8.9)$$

for $(t, x) \in R_{+} \times R^{n}$, where $A(t) = (a_{ij}(t))$, $a_{ij} \in C[R_{+}, R[\Omega, R]]$, $A^{-1}(t)$ exists and its elements belong to $M[R_{+}, R[\Omega, R_{+}]]$, and $A^{-1}(t)A'(t)$ is product-measurable and its off-diagonal elements are nonpositive for $t \in R_{+}$. Then

$$V(t, x(t, \omega), \omega) \leq R(t, \omega) \qquad \text{for} \quad t \geq t_0, \qquad (2.8.10)$$

where $R(t, \omega) = R(t, t_0, v_0, \omega)$ is the maximal solution of the random differential system

$$v'(t, \omega) = A^{-1}(t, \omega)[-A'(t, \omega)v(t, \omega) + g(t, A(t, \omega)v(t, \omega), \omega)], \qquad (2.8.11)$$
$$v(t_0, \omega) = v_0(\omega),$$

and $A(t_0, \omega)v_0(\omega) = u_0(\omega)$.

*Proof.*   We set $W(t, x, \omega) = A(t, \omega)V(t, x, \omega)$. By (2.8.9) and the nature of $A(t)$, it is easy to see that

$$D^{+}_{(2.8.1)}W(t, x, \omega) \leq g(t, W(t, x, \omega), \omega) \qquad \text{for} \quad (t, x) \in R_{+} \times R^{n}.$$

This shows that $W(t, x, \omega)$ satisfies all the hypotheses of Theorem 2.8.1, and as a consequence, we have

$$W(t, x(t, \omega), \omega) \leq r(t, \omega), \qquad t \geq t_0, \qquad (2.8.12)$$

where $r(t, \omega)$ is the maximal solution of (2.8.3) with $u_0(\omega) = A(t_0, \omega)v_0(\omega)$.

Now it is easy to observe that

$$A(t, \omega)R(t, \omega) = r(t, \omega), \qquad t \geq t_0. \tag{2.8.13}$$

Relation (2.8.12) and (2.8.13) and the properties of $A(t)$ give

$$V(t, x(t, \omega), \omega) \leq R(t, \omega) \qquad \text{for} \quad t \geq t_0,$$

which proves the theorem.

**Remark 2.8.1.** Let $B(t)$ be a product-measurable random matrix function whose elements are a.s. nonnegative locally sample Lebesgue-integrable functions on $R_+$. The random matrix function $A(t, \omega) = \exp[\int_0^t B(s, \omega) \, ds]$ is admissible in Theorem 2.8.2, provided that $A'(t, \omega)A(t, \omega) = A(t, \omega)A'(t, \omega)$.

**Remark 2.8.2.** Assume that relations (2.8.4) and (2.8.9) hold for $(t, x) \in R_+ \times B(\rho)$. If $x(t, \omega) = x(t, t_0, x_0, \omega)$ is any sample solution process of (2.8.1) such that $x_0 \in B(\rho)$ w.p. 1 and (2.8.5) holds, then (2.8.6) and (2.8.10) are valid for those values of $t \geq t_0$ for which $x(t) \in B(\rho)$ w.p. 1.

By means of the preceding comparison theorems, we are able to obtain certain upper estimates for the sample solution process of (2.8.1) in terms of the sample solution process of (2.8.3). One can also formulate comparison theorems that will give lower estimates for the sample solutions process of (2.8.1). In the following, we shall merely state a result corresponding to Theorem 2.8.1, leaving the proof to the reader.

**Theorem 2.8.3.** Assume that

(i) $g \in M[R_+ \times R^n, R[\Omega, R^m]]$ and $g(t, u, \omega)$ is a.s. sample continuous and quasi-montone nondecreasing in $u$ for fixed $t \in R_+$;

(ii) $\rho(t, \omega) = \rho(t, t_0, u_0, \omega)$ is the sample minimal solution process of (2.8.3), existing for $t \geq t_0$;

(iii) $V \in C[R_+ \times R^n, R[\Omega, R^m]]$, $V(t, x, \omega)$ is locally Lipschitzian in $x$ w.p. 1, and for $(t, x) \in R_+ \times R^n$,

$$D^+_{(2.8.1)}V(t, x, \omega) \geq g(t, V(t, x, \omega), \omega); \tag{2.8.14}$$

(iv) $x(t, \omega) = x(t, t_0, x_0, \omega)$ is any sample solution process of (2.8.1) such that

$$V(t_0, x_0, \omega) \geq u_0(\omega), \tag{2.8.15}$$

existing for $t \geq t_0$.
Then

$$V(t, x(t, \omega), \omega) \geq \rho(t, \omega) \qquad \text{for} \quad t \geq t_0.$$

**Remark 2.8.3.** Random differential systems of type (2.8.3) and (2.8.11) are called auxiliary or comparison random differential systems.

## 2.9.  SCOPE OF COMPARISON PRINCIPLE

In the following, we shall demonstrate the significance and practicability of the assumptions of the theorems of the preceding section using linear and nonlinear systems of differential equations with random coefficients.

First, we shall present some auxiliary results with regard to a random maxtrix.

Let $A(\omega) = (a_{ij}(\omega))$ be an $n \times n$ random matrix whose elements $a_{ij}(\omega)$ are random variables defined on the complete probability space $(\Omega, \mathscr{F}, P)$ into $R$.

Define

$$a(h, \omega) = \frac{1}{h} [\|I + hA(\omega)\| - 1], \tag{2.9.1}$$

where $\|\cdot\|$ is a matrix norm, $I$ is an $n \times n$ deterministic identity matrix, and $h$ is a positive real number. It is obvious that $a(h, \omega)$ is a random function defined on $R_+ \backslash \{0\}$ into $R[\Omega, R]$.

**Lemma 2.9.1.**   The random function $a(h, \omega)$ defined in (2.9.1) possesses the following properties:

    (i)   it is sample continuous in $h$;
    (ii)  it is sample increasing in $h$;
    (iii) it is bounded in $h$ for fixed $\omega \in \Omega$.

*Proof.*   First, consider the proof of (i). For any $h_0 > 0, h > 0$, and almost all $\omega \in \Omega$,

$$|a(h, \omega) - a(h_0, \omega)| = \left| \frac{1}{h} [\|I + hA(\omega)\| - 1] - \frac{1}{h_0} [\|I + h_0 A(\omega)\| - 1] \right|$$

$$\leq \frac{1}{hh_0} [[h_0\|I + hA(\omega)\| - h\|I + h_0 A(\omega)\|] + |h - h_0|]$$

$$\leq \frac{2}{hh_0} |h - h_0|.$$

This shows that

$$P\left\{ \bigcup_{h_0 \in R_+ \backslash \{0\}} \left\{ \omega : \lim_{\lambda \to 0} |a(h_0 + \lambda, \omega) - a(h_0, \omega)| \right\} \neq 0 \right\} = 0.$$

For the proof of (ii), consider $0 < h_1 < h$ such that $h_1 = \theta h$, where $0 < \theta < 1$ and $\theta = h_1/h$. Then we have

$$\|I + \theta hA(\omega)\| = \|\theta(I + hA(\omega) + (1 - \theta)I\| \leq \theta\|I + hA(\omega)\| + (1 - \theta)1,$$

and hence

$$\frac{1}{h_1}[\|I + h_1 A(\omega)\| - 1] \le \frac{1}{h}[\|I + hA(\omega)\| - 1].$$

This, together with (2.9.1), yields

$$a(h_1, \omega) \le a(h, \omega) \qquad \text{for} \quad h > h_1.$$

This proves (ii). To prove (iii), we use the properties of the norm and arrive at

$$-\|A(\omega)\| \le a(h, \omega) \le \|A(\omega)\| \qquad \text{for all} \quad h \in R_+ \backslash\{0\} \quad \text{w.p. 1.}$$

which proves (iii).

**Remark 2.9.1.** From property (i) of $a(h, \omega)$ in Lemma 2.9.1 and by the application of Theorem 1.6.2, we conclude that the stochastic process $a(h, \omega)$ is separable. Furthermore, from properties (ii) and (iii), the following limit exists:

$$\lim_{h \to 0^+} a(h, \omega) \qquad \text{exists} \quad \text{w.p. 1.} \tag{2.9.2}$$

Moreover, because of the first part of the remark, the limit in (2.9.2) is a random variable.

**Definition 2.9.1.** For any $n \times n$ random matrix,

$$\lim_{h \to 0^+} \frac{1}{h}[\|I + hA(\omega)\| - 1] \quad \text{w.p. 1} \tag{2.9.3}$$

is called the logarithmic norm of $A(\omega)$ and is denoted by $\mu(A(\omega))$.

**Remark 2.9.2.** From Remark 2.9.1, it is clear that the logarithmic norm $\mu(A(\omega))$ of an $n \times n$ random matrix is a random variable defined on $(\Omega, \mathscr{F}, P)$ with values in $R$.

**Lemma 2.9.2.** The logarithmic norm $\mu(A(\omega))$ of an $n \times n$ matrix $A(\omega)$ possesses the following properties:

$$\text{(I)} \quad \mu(\alpha A(\omega)) = \alpha \mu(A(\omega)) \qquad \text{if} \quad \alpha \ge 0, \tag{2.9.4}$$

$$\text{(II)} \quad |\mu(A(\omega))| \le \|A(\omega)\|. \tag{2.9.5}$$

Furthermore, for any $n \times n$ random matrices $A(\omega)$ and $B(\omega)$ defined on the same complete probability space $(\Omega, \mathscr{F}, P)$, the following properties hold:

$$\text{(III)} \quad \mu(A(\omega) + B(\omega)) \le \mu(A(\omega)) + \mu(B(\omega)), \tag{2.9.6}$$

$$\text{(IV)} \quad |\mu(A(\omega)) - \mu(B(\omega))| \le \|A(\omega) - B(\omega)\|. \tag{2.9.7}$$

*Proof.* The proof of properties (II) and (IV) follows from the definition of $\mu(A(\omega))$ and the properties of the norm. For $\alpha = 0$, (I) is obvious; however, for $\alpha > 0$, we note that

$$\mu(\alpha A(\omega)) = \lim_{h \to 0^+} \frac{1}{h} [\|I + \alpha A(\omega)\| - 1]$$

$$= \alpha \lim_{\alpha h \to 0^+} \frac{1}{\alpha h} [\|I + \alpha h A(\omega)\| - 1]$$

$$= \alpha u(A(\omega)).$$

This proves (I).

In order to verify the validity of (III), we observe that

$$\mu(A(\omega) + B(\omega)) = \lim_{h \to 0^+} \frac{1}{h} [\|I + h(A(\omega) + B(\omega))\| - 1]$$

$$= \lim_{2h \to 0^+} \frac{1}{2h} [\|2I + 2h(A(\omega) + B(\omega))\| - 2]$$

$$\leq \mu(A(\omega)) + \mu(B(\omega)),$$

proving (III).

**Lemma 2.9.3.** Let $\sigma(A(\omega))$ be the spectrum of an $n \times n$ random matrix $A(\omega)$. Then

$$\operatorname{Re} \lambda(\omega) \leq \mu(A(\omega)) \qquad \text{for all} \quad \lambda \in \sigma(A(\omega)) \quad \text{w.p. 1.} \qquad (2.9.8)$$

*Proof.* For any $\lambda \in \sigma(A(\omega))$, let $x$ be a corresponding characteristic vector of norm 1. Then for arbitrary $h > 0$,

$$\|(I + hA(\omega))x\| - \|x\| = |1 + h\lambda(\omega)| \|x\| - \|x\|$$
$$= |1 + h\lambda(\omega)| - 1$$
$$= h \operatorname{Re} \lambda(\omega) + O(h). \qquad (2.9.9)$$

On the other hand, from (2.9.3), we have

$$\|(I + hA(\omega))x\| - \|x\| \leq \|I + hA(\omega)\| - 1$$
$$\leq h\mu(A(\omega)) + O(h).$$

This, together with (2.9.9), yields

$$\operatorname{Re} \lambda(\omega) \leq \mu(A(\omega)) \quad \text{w.p. 1.}$$

This completes the proof of the lemma.

In the following, we give an example which shows that the logarithmic norm of a matrix depends on the norm of a matrix.

**Problem 2.9.1.** For any $n \times n$ random matrix $A(\omega)$, consider the following norms:

(MN$_1$)  $\|A(\omega)\|_E$ is the square root of the largest eigenvalue of $A^T(\omega)A(\omega)$;

(MN$_2$)  $\|A(\omega)\|_{1R} = \sup_{i \in I} \left[ \sum_{k=1}^{n} |a_{ik}(\omega)| \right]$, where $I = \{1, 2, \ldots, n\}$;

(MN$_3$)  $\|A(\omega)\|_{1C} = \sup_{k \in I} \left[ \sum_{i=1}^{n} |a_{ik}(\omega)| \right]$;

(MN$_4$)  $\|A(\omega)\|_{dC} = \sup_{k \in I} \left[ d_k^{-1} \sum_{i=1}^{n} d_i |a_{ik}(\omega)| \right]$, $d_i > 0$ and $d_i \in R$;

(MN$_5$)  $\|A(\omega)\|_Q$ is the square root of the largest eigenvalue of $Q^{-1}A^T(\omega)QA(\omega)$ for some positive definite matrix $Q$.

Then using the definition of $\mu(A(\omega))$, show that

(i)  (MN$_1$) implies

$$\mu(A(\omega)) = \text{largest eigenvalue of } \tfrac{1}{2}[A^T(\omega) + A(\omega)], \qquad (2.9.10)$$

(ii)  (MN$_2$) implies

$$\mu(A(\omega)) = \sup_{i \in I} \left[ a_{ii}(\omega) + \sum_{\substack{k=1 \\ k \neq i}}^{n} |a_{ik}(\omega)| \right], \qquad (2.9.11)$$

(iii)  (MN$_3$) implies

$$\mu(A(\omega)) = \sup_{k \in I} \left[ a_{kk}(\omega) + \sum_{\substack{i=1 \\ i \neq k}}^{n} |a_{ik}(\omega)| \right], \qquad (2.9.12)$$

(iv)  (MN$_4$) implies

$$\mu(A(\omega)) = \sup_{k \in I} \left[ a_{kk}(\omega) + d_k^{-1} \sum_{\substack{i=1 \\ i \neq k}}^{n} d_i |a_{ik}(\omega)| \right], \qquad (2.9.13)$$

(v)  (MN$_5$) implies

$$\mu(A(\omega)) = \text{largest eigenvalue of } \tfrac{1}{2}Q^{-1}[A^T(\omega)Q + QA(\omega)]. \quad (2.9.14)$$

**Remark 2.9.3.** In the preceding discussion, the elements $a_{ij}(\omega)$ of a random matrix $A(\omega)$ were in $R$. However, one can take $C$, in place of $R$, where $C$ stands for complex numbers. Then all the preceding statements remain the same except that $a_{ii}(\omega)$ in (2.9.11) and $a_{kk}(\omega)$ in (2.9.12) and (2.9.13) should be replaced by Re $a_{ii}(\omega)$ and Re $a_{kk}(\omega)$, respectively. Further, note that one can also define $\|A(\omega)\|_{dR}$ on the basis of (MN$_2$) and (MN$_4$) and can obtain $\mu(A(\omega))$.

**Remark 2.9.4.**   From Problem 2.9.1, one can easily see that the logarithmic norm is not a norm in the real sense because it can take any negative value too.

In the following, we give a problem which shows that the logarithmic norm $\mu(A(t, x, \omega))$ of a random matrix function preserves the regularity properties of $A(t, x, \omega)$.

**Problem 2.9.2.**   Let $A(t, x, \omega) = (a_{ij}(t, x, \omega))$ be an $n \times n$ random matrix process whose elements $a_{ij}$ are defined on $J \times B(z, \rho)$ into $R[\Omega, R]$. Then

(i)   $A \in M[J \times B(z, \rho), R[\Omega, R^{n^2}]]$ implies

$$\mu(A(t, x, \omega)) \in M[J \times B(z, \rho), R[\Omega, R]];$$

(ii)   $A \in C[J \times B(z, \rho), R[\Omega, R^{n^2}]]$ implies

$$\mu(A(t, x, \omega)) \in C[J \times B(z, \rho), R[\Omega, R]];$$

(iii)   $A \in M[J \times B(z, \rho), R[\Omega, R^{n^2}]]$, and $A(t, x, \omega)$ satisfies a local Lipschitz condition $x$ w.p. 1; this implies $\mu(A(t, x, \omega)) \in M[J \times B(z, \rho), R[\Omega, R]]$, and $\mu(A(t, x, \omega))$ satisfies a local Lipschitz condition in $x$ w.p. 1;

(iv)   for each $t \in J$ and almost all $\omega \in \Omega$, the sample continuity of $A(t, x, \omega)$ in $x$ implies the sample continuity of $\mu(A(t, x, \omega))$ in $x$;

(v)   the estimate

$$\|A(t, x, \omega)\| \leq K(t, \omega) \qquad \text{for} \quad (t, x) \in J \times \bar{B}(z, \rho)$$

implies

$$|\mu(A(t, x, \omega))| \leq K(t, x) \qquad \text{for} \quad (t, x) \in J \times \bar{B}(z, \rho),$$

where $K \in IB[J, R[\Omega, R_+]]$.
(*Hint*:   Use Lemma 2.9.2 for the proof.)

**Remark 2.9.5.**   We note that the numbers $d_i > 0$ in (MN$_4$) and the positive definite matrix $Q$ in (MN$_4$) can be replaced by random numbers $d_i(\omega) > 0$ w.p. 1 and the positive definite random matrix $Q(\omega)$ w.p. 1.

Now we shall discuss the scope of comparison theorems and the logarithmic norm.

Consider the system of random differential equations

$$x'(t, \omega) = A(t, \omega)x(t, \omega), \qquad x(t_0, \omega) = x_0(\omega), \qquad (2.9.15)$$

where $x \in R^n$ and $A(t, \omega)$ is an $n \times n$ random matrix whose elements $a_{ij}(t, \omega) \in M[R_+, R[\Omega, R^n]]$ and are sample Lebesgue-integrable on $R_+$. From Problem 2.9.2, it is obvious that $\mu(A(t, \omega)) \in M[R_+, R[\Omega, R]]$ and is also locally sample Lebesgue-integrable on $R_+$.

We shall prove the basic result which gives the estimate for the solution of (2.9.15) and also illustrates the significance of comparison theorems.

**Theorem 2.9.1.** Assume that the $n \times n$ random matrix function $A \in M[R_+, R[\Omega, R^{n^2}]]$, and further assume that its elements $a_{ij}(t, \omega)$ are locally sample Lebesgue-integrable on $R_+$. Then any solution $x(t, \omega)$ of (2.9.15) satisfies the relation

$$\|x(t, \omega)\| \leq \|x_0(\omega)\| \exp\left[\int_{t_0}^t \mu(A(s, \omega))\, ds\right] \qquad \text{for} \quad t \geq t_0. \quad (2.9.16)$$

*Proof.* Take $m = 1$ and $V(t, x, \omega) = \|x\|$, where $\|\cdot\|$ is any norm on $R^n$. For any $h > 0$,

$$\frac{1}{h}\left[\|x + hA(t, \omega)\| - \|x\|\right] \leq \frac{1}{h}\left[\|I + hA(t, \omega)\| \|x\| - \|x\|\right]$$

$$\leq \frac{1}{h}\left[\|I + hA(t, \omega)\| - 1\right]\|x\|.$$

This, together with the definitions of $V(t, x, \omega)$, $\mu(A(t, \omega))$, and $D^+_{(2.9.15)} V(t, x, \omega)$, gives

$$D^+_{(2.9.15)} V(t, x, \omega) \leq g(t, V(t, x, \omega), \omega) \qquad \text{for all} \quad (t, x) \in R_+ \times R^n, \quad (2.9.17)$$

where $g(t, u, \omega) = \mu(A(t, \omega))u$. From our earlier remark, it is clear that $g(t, u, \omega)$ belongs to $M[R_+ \times R, R[\Omega, R]]$, and it is sample continuous in $u$ for fixed $t \in R_+$. Furthermore, it is easy to see that the auxiliary or comparison random differential equation

$$u'(t, \omega) = \mu(A(t, \omega))u(t, \omega), \qquad u(t_0, \omega) = u_0(\omega), \quad (2.9.18)$$

has the unique solution $u(t, \omega) = u_0(\omega) \exp[\int_{t_0}^t \mu(A(s, \omega))\, ds]$ through $(t_0, u_0)$. Hence $u_0(\omega) \exp[\int_{t_0}^t \mu(A(s, \omega))\, ds]$ is also the maximal solution $r(t, \omega)$ of (2.9.18). Let $x(t, \omega)$ be any sample solution of (2.9.15) such that

$$V(t_0, x_0(\omega), \omega) = \|x_0(\omega)\| = u_0(\omega). \quad (2.9.19)$$

Thus all the hypotheses of Theorem 2.8.1 with $m = 1$ are satisfied. Hence the conclusion of Theorem 2.8.1 remains true, that is,

$$V(t, x(t, \omega), \omega) \leq r(t, \omega) \qquad \text{for} \quad t \geq t_0.$$

This, together with the definition of $V(t, x, \omega)$ and the maximal solution of (2.9.18) and (2.9.19), yields

$$\|x(t, \omega)\| \leq \|x_0(\omega)\| \exp\left[\int_{t_0}^t \mu(A(s, \omega)) \, ds\right], \qquad t \geq t_0,$$

which proves the theorem.

**Remark 2.9.6.** From Problem 2.9.1, the estimate (2.9.16) for the solution process of (2.9.15) can be represented in five different forms. Each of these forms has a certain advantage over the other. For example, the estimates corresponding to $\mu(A(t, \omega))$ in (ii)–(iv) of Problem 2.9.1 are algebraicly simple and easy to compute; however, the $\mu(A(t, \omega))$ in (i) and (v) are interesting because of the historical and analytical point of view. Further, we note that in order to get five different types of estimate for the solution process of (2.9.15) corresponding to the $\mu(A(t, \omega))$ in (i)–(v) of Problem 2.9.1, we need to specify the norm $\|\cdot\|$ of $x \in R^n$, such that $\|x\|_E = (x^T x)^{1/2}$, $\|x\|_S = \sup_{i \in I}\{|x_i|\}$, $\|x\|_C = \sum_{i=1}^n |x_i|$, $\|x\|_d = \sum_{i=1}^n d_i |x_i|$ for $d_i > 0$ and $i \in I$, and $\|x\|_Q = (x^T Q X)^{1/2}$, respectively.

Consider a particular kind of nonlinear system of random differential equations

$$x'(t, \omega) = A(t, x(t, \omega), \omega) f(x(t, \omega), \omega), \qquad x(t_0, \omega) = x_0(\omega), \quad (2.9.20)$$

where $x \in R^n$, $A(t, x, \omega) = (a_{ij}(t, x, \omega))$ is an $n \times n$ random matrix function whose elements $a_{ij} \in M[R_+ \times B(z, \rho), R[\Omega, R]]$, $a_{ij}(t, x, \omega)$ is a.s. sample continuous in $x$ for fixed $t$, $f \in R[B(z, \rho), R[\Omega, R^n]]$, $f(x, \omega) = (f_1(x, \omega), f_2(x, \omega), \ldots, f_n(x, \omega))^T$, and its sample derivative $(\partial/\partial x) f(x, \omega)$ exists and is sample continuous in $x$. Further, assume that the random functions $A(t, x, \omega)$ and $f(x, \omega)$ are smooth enough to assure the existence of a sample solution process $x(t, \omega)$ of (2.9.20), as far as $x(t) \in B(z, \rho)$, for $t \geq t_0$ w.p. 1.

**Theorem 2.9.2.** Let $A(t, x, \omega)$ and $f(x, \omega)$ be as described above. Let $V(t, x, \omega) = \|f(x, \omega)\|$, where $\|\cdot\|$ is any norm on $R^n$. Then for any solution process $x(t, \omega)$ of (2.9.20),

$$V(t, x(t, \omega), \omega) \leq V(t_0, x_0(\omega), \omega) \exp\left[\int_{t_0}^t \mu(f_x(x, \omega) A(s, x, \omega)) \, ds\right], \quad (2.9.21)$$

as far as $x(t) \in B(z, \rho)$, for $t \geq t_0$.

*Proof.* For small $h > 0$, we have

$$V(t + h, x + hA(t, x, \omega) f(x, \omega), \omega)$$
$$= \|f(x + hA(t, x, \omega) f(x, \omega), \omega)\|$$
$$= \|f(x, \omega) + hf_x(x, \omega) A(t, x, \omega) f(x, \omega) + O(h)\|,$$

which implies

$$V(t + h, x + hA(t, x, \omega)f(x, \omega), \omega) - V(t, x, \omega)$$
$$= \|f(x, \omega) + hf_x(x, \omega)A(t, x, \omega)f(x, \omega) + O(h)\| - \|f(x, \omega)\|$$
$$\le \|I + hf_x(x, \omega)A(t, x, \omega)\| \|f(x, \omega)\| - \|f(x, \omega)\| + O(h)$$
$$\le [\|I + hf_x(x, \omega)A(t, x, \omega)\| - 1]V(t, x, \omega) + O(h).$$

This, together with Definition 2.9.1 and (2.8.2), shows that

$$D^+_{(2.9.20)}V(t, x, \omega) \le \mu(f_x(x, \omega)A(t, x, \omega))V(t, x, \omega)$$

$$\text{for} \quad (t, x) \in B(z, \rho). \quad (2.9.22)$$

By (2.9.22) and Theorem 2.8.1, we then get

$$V(t, x(t, \omega), \omega) \le V(t_0, x_0(\omega), \omega) \exp\left[\int_{t_0}^{t} \mu(f_x(x, \omega)A(s, x, \omega)) \, ds\right],$$

which completes the proof of the theorem.

**Problem 2.9.3.** Assume that the hypotheses of Theorem 2.9.2 hold. Then

(i)  $(MN_1)$ in Problem 2.9.1 and $V(t, x, \omega) = (f^T(x, \omega)f(x, \omega))^{1/2}$ imply that $\mu(f_x(x, \omega)A(t, x, \omega))$ in (2.9.21) is the largest eigenvalue of

$$\tfrac{1}{2}[(f_x(x, \omega)A(t, x, \omega))^T + f_x(x, \omega)A(t, x, \omega)];$$

(ii)  $(MN_2)$ in Problem 2.9.1 and $V(t, x, \omega) = \sup_{i \in I}\{|f_i(x, \omega)|\}$ imply

$$\mu(f_x(x, \omega)A(t, x, \omega)) = \sup_{i \in I}\left[b_{ii}(t, x, \omega) + \sum_{\substack{k=1 \\ k \ne i}}^{n} |b_{ik}(t, x, \omega)|\right],$$

where $(b_{ij}(t, x, \omega)) = f_x(x, \omega)A(t, x, \omega)$;

(iii)  $(MN_3)$ in Problem 2.9.1 and $V(t, x, \omega) = \sum_{i=1}^{n}|f_i(x, \omega)|$ imply

$$\mu(f_x(x, \omega)A(t, x, \omega)) = \sup_{k \in I}\left[b_{kk}(t, x, \omega) + \sum_{\substack{i=1 \\ i \ne k}}^{n} |b_{ik}(t, x, \omega)|\right],$$

where $(b_{ij}(t, x, \omega))$ is as in (ii);

(iv)  $(MN_4)$ in Problem 2.9.1 and $V(t, x, \omega) = \sum_{i=1}^{n} d_i|f_i(x, \omega)|$ imply

$$\mu(f_x(x, \omega)A(t, x, \omega)) = \sup_{k \in I}\left[b_{kk}(t, x, \omega) + d_k^{-1} \sum_{\substack{i=1 \\ i \ne k}}^{n} d_i|b_{ik}(t, x, \omega)|\right],$$

(v)  $(MN_5)$ in Problem 2.9.1 and $V(t, x, \omega) = (f^T(x, \omega)Qf(x))^{1/2}$ imply $\mu(f_x(x, \omega)A(t, x, \omega))$ is the largest eigenvalue of

$$\tfrac{1}{2}Q^{-1}[(f_x(x, \omega)A(t, x, \omega))^TQ + Qf_x(x, \omega)A(t, x, \omega)].$$

The following problem shows the usefulness of the concept of the logarithmic norm relative to the random matrix differential equation (2.6.21).

**Problem 2.9.4.**  Show that the solution process of (2.6.21) satisfies the estimate

$$\|\Phi(t, t_0, x_0, \omega)\| \le \exp\left[\int_{t_0}^t \mu(f_x(s, x(s, \omega), \omega)) \, ds\right], \qquad t \ge t_0, \qquad (2.9.23)$$

where $\Phi(t, t_0, x_0, \omega)$ is the solution process of (2.6.21) through $(t_0, I)$ and $x(t, \omega)$ is the solution process of (2.2.1) through $(t_0, x_0)$.

The following problem shows the relationship between the solution processes of (2.2.1) and (2.6.21).

**Problem 2.9.5.**  Suppose that the hypotheses of Theorem 2.7.4 hold. Then

$$\|x(t, t_0, x_0, \omega) - x(t, t_0, y_0, \omega)\|$$
$$\le \|y_0(\omega) - x_0(\omega)\|$$
$$\times \int_0^1 \left[\exp\left[\int_{t_0}^t \mu(f_x(u, x(u, t_0, x_0 + s(y_0 - x_0), \omega), \omega)) \, du\right]\right] ds, \quad (2.9.24)$$

where for $s \in [0, 1]$, $x(u, \omega) = x(u, t_0, x_0 + s(y_0 - x_0), \omega)$ is the solution process of (2.2.1) through $(t_0, x_0 + s(y_0 - x_0))$.

The following example illustrates the comparison theorem relative to the system (2.8.1).

**Example 2.9.1.**  Suppose that (2.8.1) satisfies

$$\|x_i + h f_i(t, x, \omega)\| \le \|x_i\| + h\left(\sum_{j=1}^m a_{ij}(t, \omega)\|x_j\|\right), \qquad (2.9.25)$$

$i = 1, 2, \ldots, m$, for $(t, x) \in R_+ \times R^n$ and sufficiently small $h > 0$, where $x_i \in R^{n_i}$, $n = \sum_{i=1}^m n_i$, $a_{ij} \in M[R_+, R[\Omega, R]]$, $a_{ij}(t, \omega)$ is a.s. sample locally Lebesgue-integrable on $R_+$, and $a_{ij}(t, \omega) \ge 0$ w.p. 1 for $i \ne j$. Set

$$V(t, x, \omega) = (V_1(t, x, \omega), V_2(t, x, \omega), \ldots, V_m(t, x, \omega))^T, \qquad (2.9.26)$$

where $V_i(t, x, \omega) = \|x_i\|$, $i = 1, 2, \ldots, m$. It is obvious that

$$D_{(2.8.1)}^+ V(t, x, \omega) \le g(t, V(t, x, \omega), \omega),$$

where $g(t, u, \omega) = A(t, \omega)u$ and $A(t, \omega) = (a_{ij}(t, \omega))$ is an $m \times m$ random matrix function. Furthermore, $A(t, \omega)u$ satisfies the sample quasi-monotone nondecreasing property in $u$ for fixed $t \in R_+$. Then all hypotheses of Theorem

2.8.1 are satisfied. Hence

$$V(t, x(t, \omega), \omega) \leq r(t, \omega) \qquad \text{for} \quad t \geq t_0,$$

where $x(t, \omega)$ is any solution process of (2.8.1) and $r(t, \omega)$ is the maximal solution process of (2.8.4) with $g(t, u, \omega) = A(t, \omega)u$ and $u_0(\omega) = V(t_0, x_0(\omega), \omega)$.

**Remark 2.9.7.** We note that the most of our previous illustrations are based on the scalar version of Theorem 2.8.1. However, the real usefulness of Theorem 2.8.1 will be seen in the context of stability analysis of (2.8.1). Furthermore, this discussion of the scalar version of Theorem 2.8.1 relates to the earlier work in this field in a systematic and unified way.

## 2.10. STABILITY CONCEPTS

Let $x(t, \omega) = x(t, t_0, x_0, \omega)$ be any sample solution process of (2.8.1). Without loss of generality, we assume that $x(t, \omega) \equiv 0$ is the unique solution process of (2.8.1) through $(t_0, 0)$. Otherwise, if one finds a solution process $z(t, \omega)$ of a random algebraic equation $f(t, x, \omega) = 0$, then one can use the transformation $y = x - z(t, \omega)$ to reduce the steady state $z(t, \omega)$ of (2.8.1) to $y(t, \omega) \equiv 0$ of the transformed system

$$y'(t, \omega) = f(t, y + z(t, \omega), \omega) \equiv H(t, y(t, \omega), \omega), \qquad y(t_0, \omega) = x_0(\omega).$$

On the other hand, if one knows a solution process $z(t, \omega)$ of (2.8.1) through $(t_0, z_0)$ and is interested in knowing the stability properties of $z(t, \omega)$, then again one can use the above transformation to reduce the stability analysis of $z(t, \omega)$ to that of $y(t, \omega) \equiv 0$ of the transformed system

$$y'(t, \omega) = H(t, y(t, \omega), \omega), \qquad y(t_0, \omega) = y_0(\omega),$$

where $H(t, y, \omega) \equiv f(t, y + z(t, \omega), \omega) - f(t, z(t, \omega), \omega)$.

Now, depending on the mode of convergence in the probabilistic analysis, we shall formulate some definitions of stability.

**Definition 2.10.1.** The trivial solution of (2.8.1) is said to be

($SP_1$)   *stable in probability* if for each $\varepsilon > 0, \eta > 0, t_0 \in R_+$, there exists a positive function $\delta = \delta(t_0, \varepsilon, \eta)$ that is continuous in $t_0$ for each $\varepsilon$ and $\eta$ such that the inequality

$$P[\omega : \|x_0(\omega)\| > \delta] < \eta$$

implies

$$P[\omega : \|x(t, \omega)\| \geq \varepsilon] < \eta, \qquad t \geq t_0;$$

($SP_2$)   *uniformly stable in probability* if ($SP_1$) holds with $\delta$ independent of $t_0$;

(SP$_3$)    *asymptotically stable in probability* if it is stable in probability and
if for any $\varepsilon > 0, \eta > 0, t_0 \in R_+$, there exist numbers $\delta_0 = \delta(t_0)$ and
$T = T(t_0, \varepsilon, \eta)$ such that

$$P[\omega : \|x_0(\omega)\| > \delta_0]$$

implies

$$P[\omega : \|x(t, \omega)\| \geq \varepsilon] < \eta, \qquad t \geq t_0 + T;$$

(SP$_4$)    *uniformly asymptotically stable in probability* if (SP$_1$) and (SP$_3$)
hold with $\delta$, $\delta_0$, and $T$ independent of $t_0$;

(SS$_1$)    *stable with probability one* (or a.s. sample stable) if for each $\varepsilon > 0$,
$t_0 \in R_+$, there exists a positive function $\delta = \delta(t_0, \varepsilon)$ such that the
inequality $\|x_0(\omega)\| \leq \delta$ w.p. 1 implies

$$\|x(t, \omega)\| < \varepsilon \quad \text{w.p. 1}, \qquad t \geq t_0;$$

(SS$_2$)    *uniformly stable with probability one* if (SS$_1$) holds with $\delta$ independent of $t_0$;

(SS$_3$)    *asymptotically stable with probability one* (or a.s. sample asymptotically stable) if it is stable with probability one and if for any
$\varepsilon > 0$, $t_0 \in R_+$, there exist $\delta_0 = \delta(t_0)$ and $T = T(t_0, \varepsilon)$ such that
the inequality $\|x_0(\omega)\| \leq \delta_0$ w.p. 1 implies

$$\|x(t, \omega)\| < \varepsilon \quad \text{w.p. 1}, \qquad t \geq t_0 + T;$$

(SS$_4$)    *uniformly asymptotically stable with probability one* if (SS$_1$) and
(SS$_3$) hold with $\delta$, $\delta_0$, and $T$ independent of $t_0$;

(SM$_1$)    *stable in the pth mean*, if for each $\varepsilon > 0$, $t_0 \in R_+$, there exists a
positive function $\delta = \delta(t_0, \varepsilon)$ such that the inequality $\|x_0\|_p \leq \delta$
implies

$$\|x(t)\|_p < \varepsilon, \qquad t \geq t_0;$$

(SM$_2$)    *uniformly stable in the pth mean* if (SM$_1$) holds with $\delta$ independent
of $t_0$;

(SM$_3$)    *asymptotically stable in the pth mean* if it is stable in pth mean and
if for any $\varepsilon > 0$, $t_0 \in R_+$, there exist $\delta_0 = \delta_0(t_0)$ and $T = T(t_0, \varepsilon)$
such that the inequality $\|x_0\|_p \leq \delta_0$ implies

$$\|x(t)\|_p < \varepsilon, \qquad t \geq t_0 + T;$$

(SM$_4$)    *uniformly asymptotically stable in the pth mean* if (SM$_1$) and (SM$_3$)
hold with $\delta$, $\delta_0$, and $T$ independent of $t_0$.

**Remark 2.10.1.**    Based on Definition 2.10.1, one can formulate other
definitions of stability and boundedness. See Lakshmikantham and Leela
[59].

**Remark 2.10.2.** We note that our stability definitions are local in character as is usual. If one wants to study the global behavior of solutions such as those for boundedness and Lagrange stability, we need to take $\rho = \infty$.

For convenience, we shall now introduce certain classes of monotone functions.

**Definition 2.10.2.** A function $\phi(u)$ is said to belong to the class $\mathcal{K}$ if $\phi \in C[[0, \rho), R_+]$, $\phi(0) = 0$, and $\phi(u)$ is strictly increasing in $u$, where $0 < \rho \le \infty$.

**Definition 2.10.3.** A function $b(u)$ is said to belong to the class $\mathcal{VK}$ if $b \in C[[0, \rho), R_+]$, $b(0) = 0$, and $b(u)$ is convex and strictly increasing in $u$.

**Definition 2.10.4.** A function $a(t, u)$ is said to belong to the class $\mathcal{CK}$ if $a \in C[R_+ \times [0, \rho), R_+]$, $a(t, 0) \equiv 0$, and $a(t, u)$ is concave and strictly increasing in $u$ for each $t \in R_+$.

To use the second method of Lyapunov, which attempts to make statements about the stability properties of (2.8.1) directly by using suitable functions, we need to study the corresponding random auxiliary or comparison differential system

$$u' = g(t, u, \omega), \qquad u(t_0) = u_0(\omega), \qquad (2.10.1)$$

where $g \in M[R_+ \times R^n, R[\Omega, R^n]]$ is such that $g(t, u, \omega)$ satisfies the Carathéodory conditions in $(t, u)$ w.p. 1 and $g(t, u, \omega)$ is quasi-monotone nondecreasing in $u$ for fixed $t$ w.p. 1. We suppose that $u(t) \equiv 0$ is the solution of (2.10.1) through $(t_0, 0)$ w.p. 1.

Corresponding to the stability definitions $(SP_1)-(SP_4)$, $(SS_1)-(SS_4)$, and $(SM_1)-(SM_4)$, we designate by $(SP_1^*)-(SP_4^*)$, $(SS_1^*)-(SS_4^*)$, and $(SM_1^*)-(SM_4^*)$, the concepts concerning the stability of the equilibrium solution $u(t) \equiv 0$ of (2.10.1).

**Definition 2.10.5.** The trivial solution $u(t) \equiv 0$ of (2.10.1) is said to be

$(SP_1^*)$    *stable in probability* if given $\varepsilon > 0$, $\eta > 0$, $t_0 \in R_+$, there exists a positive function $\delta = \delta(t_0, \varepsilon, \eta)$ such that

$$P\left[\omega: \sum_{i=1}^{n} u_{i0}(\omega) > \delta\right] < \eta$$

implies

$$P\left[\omega: \sum_{i=1}^{n} u_i(t, \omega) \ge \varepsilon\right] < \eta, \qquad t \ge t_0.$$

The definitions $(SP_2^*)-(SP_4^*)$, $(SS_1^*)-(SS_4^*)$, and $(SM_1^*)-(SM_4^*)$ may be formulated similarly.

Depending on the mode of convergence in the stochastic analysis, we shall formulate some definitions of stability and boundedness of an $n \times n$ random matrix function. Let $A(t, \omega)$ be an $n \times n$ random matrix function whose $n$ column vectors are product-measurable random functions, and let $\sigma(A(t, \omega))$ denote the spectrum of $A(t, \omega)$. Let $\lambda_M(t, \omega)$ and $\lambda_m(t, \omega)$ stand for maximum and minimum eigenvalues of $A(t, \omega)$, respectively.

**Definition 2.10.6.**    An $n \times n$ random matrix function $A(t, \omega)$ is said to be

(i)  *P-stable in probability* for any $t_0 \in R_+$ if it satisfies the relation

$$\limsup_{t \to \infty} \left[ P\left\{ \omega : \int_{t_0}^t \lambda_M(s, \omega)\, ds \geq 0 \right\} \right] = 0;$$

(ii)  *P-stable with probability one* if for $t_0 \in R_+$, there exists a positive number $a = a(t_0)$ such that

$$P\left\{ \omega : \limsup_{t \to \infty} \left[ \frac{1}{t - t_0} \int_{t_0}^t \lambda_M(s, \omega)\, ds \right] \leq -a(t_0) \right\} = 1;$$

(iii)  *P-bounded in probability* if for any $\eta > 0$, $t_0 \in R_+$, there exist positive numbers $k = k(t_0)$ and $T = T(t_0, \eta)$ such that

$$P\left\{ \omega : \int_{t_0}^t \lambda_M(s, \omega)\, ds \geq k \right\} < \eta, \qquad t \geq t_0 + T;$$

(iv)  *P-bounded with probability one* if for $t_0 \in R_+$, there exist positive numbers $K = K(t_0)$ and $T = T(t_0)$ such that

$$P\left\{ \omega : \int_{t_0}^t \lambda_M(s, \omega)\, ds \leq K,\, t \geq t_0 + T \right\} = 1.$$

## 2.11.  STABILITY IN PROBABILITY

First we shall use the second method of Lyapunov to derive the stability results for (2.8.1) by recalling the definition of the random function $D_{(2.8.1)}^+ V(t, x, \omega)$ in (2.8.2).

**Theorem 2.11.1.**    Assume that

(i)  $g \in M[R_+ \times R^m, R[\Omega, R^m]]$ and $g(t, u, \omega)$ is sample continuous and quasi-monotone nondecreasing in $u$ for fixed $t \in R_+$,

(ii)  $V \in C[R_+ \times B(\rho), R[\Omega, R^m]]$, satisfies a local Lipschitz condition in $x$ w.p. 1, and for $(t, x) \in R_+ \times B(\rho)$,

$$D_{(2.8.1)}^+ V(t, x, \omega) \leq g(t, V(t, x, \omega), \omega); \tag{2.11.1}$$

(iii)   for $(t, x) \in R_+ \times B(\rho)$,

$$b(\|x\|) \le \sum_{i=1}^{m} V_i(t, x, \omega) \le a(t, \|x\|, \omega), \qquad (2.11.2)$$

where $b \in \mathcal{K}$, $a(t, \cdot, \omega) \in \mathcal{K}$ and $a \in C[R_+ \times R_+, R[\Omega, R_+]]$.
Then

$$(SP_1^*) \qquad \text{implies} \qquad (SP_1).$$

*Proof.*   Let $\eta > 0$, $0 < \varepsilon < \rho$, and $t_0 \in R_+$ be given. Assume that $(SP_1^*)$
holds. Then given $b(\varepsilon) > 0, \eta > 0$, and $t_0 \in R_+$, there exists a positive function
$\delta_1 = \delta_1(t_0, \varepsilon, \eta)$ such that

$$P\left\{\omega: \sum_{i=1}^{m} u_i(t, t_0, u_0, \omega) \ge b(\varepsilon)\right\} < \eta, \qquad t \ge t_0, \qquad (2.11.3)$$

provided that

$$P\left\{\omega: \sum_{i=1}^{m} u_{i0}(\omega) > \delta_1\right\} < \eta. \qquad (2.11.4)$$

Let us choose $u_0 = (u_{10}, u_{20}, \ldots, u_{mn})^T$ so that $V(t_0, x_0(\omega), \omega) \le u_0(\omega)$ and

$$\sum_{i=1}^{m} u_{i0}(\omega) = a(t_0, \|x_0(\omega)\|, \omega) \qquad \text{for} \quad x_0 \in B(\rho). \qquad (2.11.5)$$

Since $a(t_0, \cdot, \omega) \in \mathcal{K}$, we can find a $\delta = \delta(t_0, \varepsilon, \eta)$ such that

$$P\{\omega: a(t_0, \|x_0(\omega)\|, \omega) > \delta_1\} = P\{\omega: \|x_0(\omega)\| > \delta\}. \qquad (2.11.6)$$

Now we claim that $(SP_1)$ holds. Suppose that this claim is false. Then
there would exist a sample solution process $x(t, \omega)$ of $(2.8.1)$ with $P\{\omega: \|x_0(\omega)\| > \delta\} < \eta$ and a $t_1 > t_0$ such that

$$P\{\omega: \|x(t_1, \omega)\| \ge \varepsilon\} = \eta \qquad \text{and} \qquad x(t) \in B(\rho) \qquad \text{for} \quad t \in [t_0, t_1].$$
$$(2.11.7)$$

On the other hand, by Theorem 2.8.1, the inequality

$$V(t, x(t, \omega), \omega) \le r(t, t_0, u_0, \omega) \qquad (2.11.8)$$

is valid so long as $x(t, \omega) \in B(\rho)$. From $(2.11.2)$ and $(2.11.8)$, we have

$$b(\|x(t, \omega)\|) \le \sum_{i=1}^{m} V_i(t, x(t, \omega), \omega) \le \sum_{i=1}^{m} r_i(t, t_0, u_0, \omega). \qquad (2.11.9)$$

Relations (2.11.3), (2.11.7), and (2.11.9) lead us to the contradiction

$$\eta = P\{\omega:\|x(t_1,\omega)\| \geq \varepsilon\} = P\{\omega:b(\|x(t_1,\omega)\|) \geq b(\varepsilon)\}$$

$$\leq P\left\{\omega: \sum_{i=1}^{m} r_i(t_1,t_0,u_0,\omega) \geq b(\varepsilon)\right\} < \eta,$$

which proves (SP$_1$). The proof of the theorem is complete.

**Example 2.11.1.** Let $\mu(A(t,\omega))$ be the logarithmic norm of $A(t,\omega)$ in (2.9.15). Then for any $\eta > 0$, $t_0 \in R_+$, there exist numbers $k(t_0) = k > 0$ and $T(t_0,\eta) = T > 0$ such that

$$P\left\{\omega: \int_{t_0}^t \mu(A(s,\omega))\, ds \geq k\right\} < \eta, \qquad t \geq t_0 + T, \qquad (2.11.10)$$

implies (SP$_1$) relative to (2.9.15).

*Proof.* From the proof of Theorem 2.9.1, it is obvious that the functions $g(t,u,\omega) = \mu(A(t,\omega))u$ and $V(t,x,\omega) = \|x\|$ satisfy hypotheses (i)–(iii) of Theorem 2.11.1. Now it remains to show that (SP$_1^*$) holds. To show this, let $\varepsilon > 0$, $\eta > 0$, and $t_0 \in R_+$ be given. From the hypothesis of the example, there exist numbers $T(t_0,\eta) = T$ and $k > 0$ such that (2.11.10) holds. Note that

$$u(t,t,u,\omega) = u_0(\omega) \exp\left[\int_{t_0}^t \mu(A(s,\omega))\, ds\right], \qquad (2.11.11)$$

where $u(t,t_0,u_0,\omega)$ is the solution process of the scalar version of (2.10.1) with $g(t,u,\omega) = \mu(A(t,\omega))u$. Choose $\delta_1(t_0,\varepsilon) = \delta_1 = \varepsilon/e^{k(t_0)}$ so that

$$u_0(\omega) \leq \delta_1 \quad \text{w.p. 1}. \qquad (2.11.12)$$

From (2.11.11) and (2.11.12), we obtain

$$P\{\omega:u(t,\omega) \geq \varepsilon\} \leq P\left[\omega:0 \leq \left[-k + \int_{t_0}^t \mu(A(s,\omega))\, ds\right]\right].$$

This, together with (2.11.10), yields

$$P\{\omega:u(t,\omega) \geq \varepsilon\} < \eta, \qquad t \geq t_0 + T. \qquad (2.11.13)$$

First, we choose a positive number $K_0$ such that

$$P\left\{\omega: \int_{t_0}^{t_0+T} |\mu(A(s,\omega))|\, ds \geq K_0\right\} < \eta,$$

and then we choose $u_0(\omega)$ such that $u_0(\omega) \leq \delta_2$, where $\delta_2 = \varepsilon/e^{K_0}$. For

$t \in [t_0, t_0 + T]$, relation (2.11.11) gives

$$P\{\omega : u(t, \omega) \geq \varepsilon\} \leq P\left\{\omega : 1 \leq \exp\left[-K_0 + \int_{t_0}^{t} \mu(A(s, \omega)) \, ds\right]\right\}$$

$$\leq P\left\{\omega : 0 \leq \left[-K_0 + \int_{t_0}^{t} \mu(A(s, \omega)) \, ds\right]\right\} < \eta. \quad (2.11.14)$$

By (2.11.13) and (2.11.14), we get

$$P\{\omega : u(t, \omega) \geq \varepsilon\} < \eta, \qquad t \geq t_0,$$

whenever $P\{\omega : u_0(\omega) > \delta\} < \eta$, where $\delta = \min(\delta_1, \delta_2)$. This proves (SP$_1^*$), and hence by Theorem 2.11.1, (SP$_1$) holds.

**Remark 2.11.1.** Consider Problem 2.9.1. Then stability condition (2.11.10) relative to the norms (MN$_1$)–(MN$_5$) reduces to the following respective stability conditions:

the random matrix $\frac{1}{2}[A^T(t, \omega) + A(t, \omega)]$ is $P$-bounded in probability,

$$(2.11.15)$$

$$P\left\{\omega : \int_{t_0}^{t} \sup_{i \in I} \left[a_{ii}(s, \omega) + \sum_{\substack{k=1 \\ k \neq i}}^{n} |a_{ik}(s, \omega)|\right] ds \geq k(t_0)\right\} < \eta,$$

$$t \geq t_0 + T, \quad (2.11.16)$$

$$P\left\{\omega : \int_{t_0}^{t} \sup_{k \in I} \left[a_{kk}(s, \omega) + \sum_{\substack{i=1 \\ i \neq k}}^{n} |a_{ik}(s, \omega)|\right] ds \geq k(t_0)\right\} < \eta,$$

$$t \geq t_0 + T, \quad (2.11.17)$$

$$P\left\{\omega : \int_{t_0}^{t} \sup_{k \in I} \left[a_{kk}(s, \omega) + d_k^{-1} \sum_{\substack{i=1 \\ i \neq k}}^{n} d_i |a_{ik}(s, \omega)|\right] ds \geq k(t_0)\right\} < \eta,$$

$$t \geq t_0 + T, \quad (2.11.18)$$

and

the random matrix $\frac{1}{2}Q^{-1}[A^T(t, \omega)Q + QA(t, \omega)]$ is $P$-bounded in probability.

$$(2.11.19)$$

**Remark 2.11.2.** The function $g(t, u, \omega) = A(t, \omega)u$ is admissible in Theorem 2.11.1, where the off-diagonal elements $a_{ij}(t, \omega) \geq 0$, $i \neq j$, if $\mu(A(t, \omega))$ satisfies (2.11.10). Clearly, the function $g(t, u, \omega) \equiv 0$ w.p. 1 is also admissible.

**Theorem 2.11.2.** Let the hypotheses of Theorem 2.11.1 be satisfied with $a(t, r, \omega) = a(r, \omega)$. Then

$$(\text{SP}_2^*) \quad \text{implies} \quad (\text{SP}_2).$$

*Proof.* By following the proof of Theorem 2.11.1, it is easy to see that $\delta$ does not depend on $t_0$. This is because of the fact that $a(t, r, \omega) = a(r, \omega)$ and the uniform stability in probability of (2.10.1).

**Problem 2.11.1.** Define the notions of boundedness in probability and give a set of sufficient conditions so that one of these concepts holds.

**Theorem 2.11.3.** Let the hypotheses of Theorem 2.11.1 be satisfied. Then

$$(\text{SP}_3^*) \quad \text{implies} \quad (\text{SP}_3).$$

*Proof.* We note that $(\text{SP}_3^*)$ implies $(\text{SP}_1^*)$. Hence, by Theorem 2.11.1, the trivial solution of (2.8.1) is stable in probability. We fix $\varepsilon = \rho$, $\eta = \eta_0 < 1$, and designate by $\delta_0^*$ the number $\delta(t_0, \rho, \eta_0)$. To show that $(\text{SP}_3)$ holds, it is enough to prove that for any $0 < \varepsilon < \rho$, $0 < \eta < \eta_0$, and $t_0 \in R_+$, there exist positive numbers $\delta_0 = \delta(t_0)$ and $T = T(t_0, \varepsilon, \eta)$ such that $P\{\omega : \|x_0(\omega)\| > \delta_0\} > \eta$ implies

$$P\{\omega : \|x(t, \omega)\| \geq \varepsilon\} < \eta, \qquad t \geq t_0 + T. \tag{2.11.20}$$

Assume that $(\text{SP}_3^*)$ holds. Then given $b(\varepsilon) > 0$, $\eta > 0$, and $t_0 \in R_+$, there exist numbers $\delta^0(t_0) = \delta^0$ and $T(t_0, \varepsilon, \eta) = T > 0$ such that

$$P\left\{\omega : \sum_{i=1}^{m} u_i(t, t_0, u_0, \omega) \geq b(\varepsilon)\right\} < \eta, \qquad t \geq t_0 + T, \tag{2.11.21}$$

whenever

$$P\left\{\omega : \sum_{i=1}^{m} u_{i0}(\omega) > \delta^0\right\} < \eta.$$

As before, we choose $u_0$ so that (2.11.5) holds and choose $\delta_0^{**}(t_0) = \delta_0^{**}$ such that

$$P\{\omega : a(t_0, \|x_0(\omega)\|, \omega) > \delta^0\} = P\{\omega : \|x_0(\omega)\| > \delta_0^{**}\}.$$

Then choose $\delta_0 = \min(\delta_0^*, \delta_0^{**})$. With this $\delta_0$, we claim that (2.11.20) holds. Otherwise, there exists a sequence $\{t_n\}$, $t_n \geq t_0 + T$, $t_n \to \infty$ as $n \to \infty$, such that for some solution process (2.8.1) satisfying $P\{\omega : \|x_0(\omega)\| > \delta_0\} < \eta$, we have $P\{\omega : \|x(t_n, \omega)\| \geq \varepsilon\} = \eta$, $t_n \geq t_0 + T$. This, together with (2.11.9) and (2.11.21), will establish the validity of (2.11.20). This completes the proof of the theorem.

**Example 2.11.2.** Let $\mu(A(t,\omega))$ be the logarithmic norm of $A(t,\omega)$ in (2.9.15). Then for any $\eta > 0$ and $t_0 \in R_+$, there exists a number $T(t_0, \eta) = T > 0$ such that

$$P\left\{\omega: \int_{t_0}^t \mu(A(s,\omega))\, ds \geq 0\right\} < \eta, \qquad t \geq t_0 + T, \qquad (2.11.22)$$

implies (SP$_3$) of (2.9.15).

*Proof.* From the proof of Example 2.11.1, the functions $V(t,x,\omega) = \|x\|$ and $g(t,u,\omega) = \mu(A(t,\omega))u$ satisfy the hypotheses of Theorem 2.11.3. Now it remains to show that (SP$_3^*$) holds. Let $\varepsilon > 0$, $\eta > 0$, and $t_0 \in R_+$ be given. Note that

$$P\left\{\omega: \int_{t_0}^t \mu(A(s,\omega))\, ds \geq \varepsilon\right\} \leq P\left\{\omega: \int_{t_0}^t \mu(A(s,\omega))\, ds \geq 0\right\}, \qquad t \geq t_0 + T.$$

From this and following the proof of Example 2.11.1, we can conclude that (SP$_1^*$) holds for (2.9.15). From (2.11.22), it follows that

$$\lim_{t \to \infty} P\left\{\omega: \int_{t_0}^t \mu(A(s,\omega))\, ds \geq 0\right\} = 0.$$

Hence we conclude that (SP$_3^*$) holds for (2.9.15).

**Remark 2.11.3.** For Problem 2.9.1, the stability condition (2.11.22) corresponding to the norms (MN$_1$)–(MN$_5$) reduces to the following respective stability conditions:

the random matrix $\frac{1}{2}[A^{\mathsf{T}}(t,\omega) + A(t,\omega)]$ is $P$ stable in probability,

$$(2.11.23)$$

$$P\left\{\omega: \int_{t_0}^t \sup_{i \in I}\left[a_{ii}(s,\omega) + \sum_{\substack{k=1 \\ k \neq i}}^n |a_{ik}(s,\omega)|\right] ds \geq 0\right\} < \eta,$$

$$t \geq t_0 + T, \quad (2.11.24)$$

$$P\left\{\omega: \int_{t_0}^t \sup_{k \in I}\left[a_{kk}(s,\omega) + \sum_{\substack{i=1 \\ i \neq k}}^n |a_{ik}(s,\omega)|\right] ds \geq 0\right\} < \eta,$$

$$t \geq t_0 + T, \quad (2.11.25)$$

$$P\left\{\omega: \int_{t_0}^t \sup_{k \in I}\left[a_{kk}(s,\omega) + d_k^{-1}\sum_{\substack{i=1 \\ i \neq k}}^n d_i |a_{ik}(s,\omega)|\right] ds \geq 0\right\} < \eta,$$

$$t \geq t_0 + T, \quad (2.11.26)$$

and

the random matrix $\frac{1}{2}Q^{-1}[A^{\mathrm{T}}(t, \omega)Q + QA(t, \omega)]$ is $P$-stable in probability.

$$(2.11.27)$$

**Theorem 2.11.4.**   Assume that the hypotheses of Theorem 2.11.2 hold. Then

$$(\mathrm{SP}_4^*)   \quad \text{implies}   \quad (\mathrm{SP}_4).$$

*Proof.*   The proof follows from the proofs of Theorems 2.11.2 and 2.11.3.

In certain cases, results that are based on the comparison theorem (Theorem 2.8.2) are useful in discussing the stability properties of (2.8.1). We shall simply state the result corresponding to Theorem 2.11.1, leaving its proof as an exercise.

**Theorem 2.11.5.**   Assume that the hypotheses of Theorem 2.11.1 hold except that relation (2.11.1) is replaced by

$$A(t)D_{(2.8.1)}^{+}V(t, x, \omega) + A'(t)V(t, x, \omega) \le g(t, A(t)V(t, x, \omega), \omega),$$

where $A(t) = (a_{ij}(t))$, $a_{ij} \in C[R_+, R[(\Omega, F, P), R]]$, $A^{-1}(t)$ exists and its elements belong to $M[R_+, R[\Omega, R_+]]$ w.p. 1, and $A^{-1}(t)A'(t)$ is product-measurable with off-diagonal elements $a_{ij}$, $i \ne j$, nonnegative for $t \in R_+$. Then the $(\mathrm{SP}_1^*)$ property of the trivial solution of (2.8.11) implies $(\mathrm{SP}_1)$.

Now we shall use the method of variation of parameters to derive some stability results in probability for (2.8.1) and its nonlinear perturbations (2.7.9).

**Theorem 2.11.6.**   Assume that

(i)   the hypotheses of Theorem 2.7.4 hold;

(ii)   $f(t, 0, \omega) \equiv 0$ w.p. 1, $t \in R_+$;

(iii)   there exists a random function $\alpha \in C[R_+, R[\Omega, R]]$ such that for each $t \in R_+$, $x_0(\omega)$ in (2.8.1) and $\alpha(t, \omega)$ are independent random variables and

$$\mu(f_x(t, x, \omega)) \le \alpha(t, \omega),   \qquad (t, x) \in R_+ \times B(\rho). \qquad (2.11.28)$$

Then for any $\eta > 0$ and $t_0 \in R_+$, there exists a positive number $T = T(t_0, \eta)$ such that

$$P\left\{\omega : \int_{t_0}^{t} \alpha(s, \omega)\, ds \ge 0\right\} < \eta, \qquad t \ge t_0 + T, \qquad (2.11.29)$$

implies $(\mathrm{SP}_3)$ of (2.8.1).

*Proof.* From hypothesis (ii), we have $x(t, t_0, 0) \equiv 0$ w.p. 1, where $x(t, t_0, x_0, \omega)$ is a solution process of (2.8.1). This, together with Theorem 2.7.4 and Problem 2.9.5 and relation (2.11.28), yields

$$\|x(t, t_0, x_0, \omega)\| \le \exp\left[\int_{t_0}^t \alpha(s, \omega)\, ds\right]\|x_0(\omega)\|, \qquad t \ge t_0. \quad (2.11.30)$$

Now the rest of the proof follows by using the arguments in the proofs of Examples 2.11.1 and 2.11.2.

**Theorem 2.11.7.** Assume that

(i)   the hypotheses of Theorem 2.7.3 hold,
(ii)  $f(t, 0, \omega) \equiv F(s, 0, \omega) \equiv 0$ w.p. 1, $t \in R_+$,
(iii) hypothesis (iii) of Theorem 2.11.6 holds, and
(iv)  $\|F(t, x, \omega)\| = o(\|x\|)$ as $x \to 0$ uniformly in $t$ w.p. 1.

Then relation (2.11.29) implies $(SP_3)$ of (2.7.9).

*Proof.* Let $\varepsilon > 0$ be sufficiently small. Then relation (2.11.29) implies

$$P\left\{\omega : \varepsilon(t - t_0) + \int_{t_0}^t \alpha(s, \omega)\, ds \ge 0\right\} < \eta, \qquad t \ge t_0 + T. \quad (2.11.31)$$

Hence one can choose a positive constant $K$ such that

$$P\left\{\omega : \left[\varepsilon(t - t_0) + \int_{t_0}^t \alpha(s, \omega)\, ds\right] \ge K\right\} < \eta, \qquad t \ge t_0. \quad (2.11.32)$$

With the foregoing $\varepsilon$ and because of hypothesis (iv), there exists a $\delta > 0$ such that $\|x\| < \delta$ implies $\|F(t, x, \omega)\| < \varepsilon\|x\|$. Now for $\|x_0\| < \delta/e^K$ and from Theorem 2.9.5, (2.7.14), and (2.11.30), we obtain

$$\|y(t, t_0, x_0(\omega), \omega)\| \le \|x_0(\omega)\| \exp\left[\int_{t_0}^t \alpha(s, \omega)\, ds\right]$$
$$+ \varepsilon \int_{t_0}^t \exp\left[\int_s^t \alpha(u, \omega)\, du\right]\|y(s, t_0, x_0(\omega), \omega)\|\, ds,$$

$$(2.11.33)$$

so long as $\|y(t, t_0, x_0(\omega), \omega)\| < \delta$. Multiplying both sides of (2.11.33) by $\exp[-\int_{t_0}^t \alpha(s, \omega)\, ds]$ and applying the scalar version of Theorem 1.7.2, we get

$$\|y(t, t_0, x_0(\omega), \omega)\| \le \|x_0(\omega)\| \exp\left[\varepsilon(t - t_0) + \int_{t_0}^t \alpha(s, \omega)\, ds\right], \quad (2.11.34)$$

so long as $\|y(t, t_0, x_0(\omega), \omega)\| < \delta$. Now (2.11.34) shows that
$$\|y(t, t_0, x_0(\omega), \omega)\| < \delta \qquad \text{for all} \quad t \ge t_0.$$

Otherwise for any $\eta > 0$, there exists a $T$ such that

$$P\{\omega: \|y(T, t_0, x_0(\omega), \omega)\| \geq \delta\} = \eta.$$

Then from (2.11.34) and (2.11.32), we get the contradiction

$$\eta = P\{\omega: \|y(T, t_0, x_0(\omega), \omega)\| \geq \delta\}$$

$$\leq P\left\{\omega: \exp\left[-K + \varepsilon(t - t_0) + \int_{t_0}^t \alpha(s, \omega)\, ds\right] \geq 1\right\}$$

$$\leq P\left\{\omega: \varepsilon(t - t_0) + \int_{t_0}^t \alpha(s, \omega)\, ds \geq K\right\} < \eta,$$

which proves our claim. Thus (2.11.34) holds for all $t \geq t_0$, and this, together with (2.11.31), yields the desired conclusion.

### 2.12.   STABILITY WITH PROBABILITY ONE

In this section, we shall present some stability criteria that assure the stability with probability one or almost-surely sample stability properties of the trivial solution of (2.8.1). Furthermore, some illustrations are given to show that the stability conditions are connected with the statistical properties of the random rate function of the differential equations. An example is also worked out to exhibit the advantage of the use of the vector Lyapunov function over the single Lyapunov function.

First, we shall prove some results for stability with probability one in the context of random differential inequalities and Lyapunov functions. We then consider results for stability with probability one in the framework of the method of variation of parameters.

**Theorem 2.12.1.**   Assume that conditions (i)–(iii) of Theorem 2.11.1 hold. Then

$$(SS_1^*) \qquad \text{implies} \qquad (SS_1).$$

*Proof.*   Let $0 < \varepsilon < \rho$ and $t_0 \in R_+$ be given. Assume that $(SS_1^*)$ holds. Then given $b(\varepsilon) > 0$ and $t_0 \in R_+$, we can find a positive number $\delta_1 = \delta(t_0, \varepsilon)$ such that

$$\sum_{i=1}^m u_{i0}(\omega) \leq \delta_1 \tag{2.12.1}$$

implies

$$\sum_{i=1}^m u_i(t, t_0, u_0, \omega) < b(\varepsilon), \qquad t \geq t_0. \tag{2.12.2}$$

Let us choose $u_0(\omega)$ such that $V(t_0, x_0, \omega) \leq u_0(\omega)$ and

$$\sum_{i=1}^{m} u_{i0}(\omega) = a(t_0, \|x_0(\omega)\|, \omega). \qquad (2.12.3)$$

Since $a(t, \cdot, \omega) \in \mathcal{K}$ and $a \in C[R_+ \times R_+, R[\Omega, R_+]]$, we can find $\delta = \delta(t_0, \varepsilon) > 0$ such that

$$a(t_0, \|x_0(\omega)\|, \omega) \leq \delta_1 \quad \text{and} \quad \|x_0(\omega)\| \leq \delta \qquad (2.12.4)$$

hold simultaneously. We claim that if $\|x_0(\omega)\| \leq \delta$, then

$$\|x(t, t_0, x_0, \omega)\| < \varepsilon, \qquad t \geq t_0.$$

Suppose that this claim is false. Then there would exist a sample solution process $x(t, t_0, x_0, \omega)$ with $\|x_0(\omega)\| \leq \delta$, $\Omega_1 \subset \Omega$, $P(\Omega_1) > 0$, and $t_1 > t_0$ such that

$$\|x(t_1, \omega)\| = \varepsilon, \quad \omega \in \Omega_1, \quad \text{and} \quad \|x(t, \omega)\| < \varepsilon, \qquad t \in [t_0, t_1). \quad (2.12.5)$$

This implies that $x(t) \in B(\rho)$ for $t \in [t_0, t_1)$, and hence, from Theorem 2.8.1, we have

$$V(t, x(t, \omega), \omega) \leq r(t, t_0, u_0, \omega) \qquad \text{for} \quad t \in [t_0, t_1). \qquad (2.12.6)$$

It therefore follows that

$$b(\|x(t, \omega)\|) \leq \sum_{i=1}^{m} V_i(t, x(t, \omega), \omega) \leq \sum_{i=1}^{m} r_i(t, \omega) \qquad \text{for} \quad t \in [t_0, t_1). \quad (2.12.7)$$

Relations (2.12.2), (2.12.5), and (2.12.7) and the continuity of the functions involved lead us to the contradiction

$$b(\varepsilon) \leq \sum_{i=1}^{m} V_i(t_1, x(t_1, \omega), \omega) \leq \sum_{i=1}^{m} r_i(t_1, \omega) < b(\varepsilon), \qquad \omega \in \Omega_1.$$

This establishes the fact that $(SS_1)$ holds. The proof of the theorem is complete.

**Example 2.12.1.** Let $\mu(A(t, \omega))$ be the logarithmic norm of $A(t, \omega)$ in (2.9.15). Then for $t_0 \in R_+$, there exist deterministic positive numbers $K = K(t_0)$ and $T = T(t_0)$ such that

$$P\left\{\omega : \int_{t_0}^{t} |\mu(A(s, \omega))| \, ds \leq K, t \geq t_0 + T\right\} = 1 \qquad (2.12.8)$$

implies $(SS_1)$ of (2.9.15).

*Proof.* By following the argument used in the proof of Example 2.11.1, the conclusion of the example remains true, provided that $(SS_1^*)$ holds for

(2.10.1) with $m = 1$ and $g(t, u, \omega) = \mu(A(t, \omega))u$. Observe that the sample Lebesgue integral $\int_{t_0}^t \mu(A(s, \omega)) \, ds$ is sample absolutely continuous. Therefore, from (2.12.8) and (2.9.16), the conclusion of the example remains true.

**Remark 2.12.1.** From Problem 2.9.1, stability condition (2.12.8) corresponding to the norms $(MN_1)$–$(MN_5)$ reduces to the following respective stability conditions:

the random matrix $\frac{1}{2}[A^{\mathrm{T}}(t, \omega) + A(t, \omega)]$ is $P$-bounded w.p. 1,    (2.12.9)

$$P\left\{\omega: \int_{t_0}^t \left|\sup_{i \in I}\left[a_{ii}(s, \omega) + \sum_{\substack{k=1 \\ k \neq i}}^n |a_{ik}(s, \omega)|\right]\right| ds \leq K(t_0),\right.$$

$$\left. t \geq t_0 + T \right\} = 1, \tag{2.12.10}$$

$$P\left\{\omega: \int_{t_0}^t \left|\sup_{k \in I}\left[a_{kk}(s, \omega) + \sum_{\substack{i=1 \\ i \neq k}}^n |a_{ik}(s, \omega)|\right]\right| ds \geq K(t_0),\right.$$

$$\left. t \geq t_0 + T \right\} = 1, \tag{2.12.11}$$

$$P\left\{\omega: \int_{t_0}^t \left|\sup_{k \in I}\left[a_{kk}(s, \omega) + d_k^{-1} \sum_{\substack{i=1 \\ i \neq k}}^n d_i |a_{ik}(s, \omega)|\right]\right| ds \leq K(t_0),\right.$$

$$\left. t \geq t_0 + T \right\} = 1, \tag{2.12.12}$$

and

the random matrix $\frac{1}{2}Q^{-1}[A^{\mathrm{T}}(t, \omega)Q + QA(t, \omega)]$ is $P$-bounded w.p. 1.

$$\tag{2.12.13}$$

Based on the proofs of Theorems 2.11.1, 2.11.2, and 2.12.1, it is not difficult to construct the proofs of the rest of the notions of stability w.p. 1. We shall incorporate such results in the following problem, leaving the details to the reader.

**Problem 2.12.1.** Under the hypotheses of Theorem 2.11.1, show that $(SS_3^*)$ implies $(SS_3)$. If the assumptions of Theorem 2.11.2 hold, show that $(SS_2^*)$ implies $(SS_2)$ and $(SS_4^*)$ implies $(SS_4)$.

**Problem 2.12.2.** Let $\mu(A(t, \omega))$ be the logarithmic norm of $A(t, \omega)$ in (2.9.15). Then for $t_0 \in R_+$, there exists a deterministic positive number

$a = a(t_0) > 0$ such that

$$P\left\{\omega:\limsup_{t\to\infty}\left[\frac{1}{t - t_0}\int_{t_0}^t \mu(A(s, \omega))\,ds\right] \le -a(t_0)\right\} = 1 \quad (2.12.14)$$

implies $(SS_3)$ relative to (2.9.15).

**Remark 2.12.2.** By Problem 2.9.1, stability condition (2.12.14) corresponding to the norms $(MN_1)$–$(MN_5)$ reduces to the following respective stability conditions:

the random matrix $\frac{1}{2}[A^{T}(t, \omega) + A(t, \omega)]$ is $P$-stable w.p. 1,  (2.12.15)

$$P\left\{\omega:\limsup_{t\to\infty}\left[\frac{1}{t - t_0}\int_{t_0}^t \sup_{i\in I}\left[a_{ii}(s, \omega) + \sum_{\substack{k=1\\k\ne i}}^n |a_{ik}(s, \omega)|\right]ds\right]\right.$$
$$\left.\le -a(t_0)\right\} = 1, \qquad (2.12.16)$$

$$P\left\{\omega:\limsup_{t\to\infty}\left[\frac{1}{t - t_0}\int_{t_0}^t \sup_{k\in I}\left[a_{kk}(s, \omega) + \sum_{\substack{i=1\\i\ne k}}^n |a_{ik}(s, \omega)|\right]ds\right]\right.$$
$$\left.\le -a(t_0)\right\} = 1, \qquad (2.12.17)$$

$$P\left\{\omega:\limsup_{t\to\infty}\left[\frac{1}{t - t_0}\int_{t_0}^t \sup_{k\in I}\left[a_{kk}(s, \omega) + d_k^{-1}\sum_{\substack{i=1\\i\ne k}}^n d_i|a_{ik}(s, \omega)|\right]ds\right]\right.$$
$$\left.\le -a(t_0)\right\} = 1, \qquad (2.12.18)$$

and

the random matrix $\frac{1}{2}Q^{-1}[A^{T}(t, \omega)Q + QA(t, \omega)]$ is $P$-stable w.p. 1.  (2.12.19)

Let us now discuss an example that shows a certain gain over various stability conditions.

Consider the special type of random differential system

$$x'(t, \omega) = A(\omega)x + B(t, \omega)x, \qquad x(t_0, \omega) = x_0(\omega), \qquad (2.12.20)$$

where $x \in R^n$, $\|A(\omega)\| \le K$ w.p. 1, $E[\|B(t, \omega)\|] < \infty$ for $t \in R_+$, and the elements $b_{ij}(t, \omega)$ of the random matrix function $B(t, \omega) = (b_{ij}(t, \omega))$ are product-measurable. Under these conditions it is obvious that the initial-value problem (2.12.20) has a unique sample solution for $t \ge t_0$.

In the following, we shall only discuss (SS$_4$) of (2.12.20). However, other properties (SS$_1$) and (SS$_3$) can also be discussed analogously, which we leave as exercises.

Set $A(t, \omega)x = [A(\omega) + B(t, \omega)]x = A(\omega)x + B(t, \omega)x$. This, together with (2.12.20) and the stability condition (2.12.14) in Problem 2.12.2, for (2.12.20), yields

$$P\left\{\omega:\limsup_{t\to\infty}\left[\frac{1}{t-t_0}\int_{t_0}^{t}\mu(A(\omega) + B(s, \omega))\,ds\right]\le -a\right\} = 1, \quad (2.12.21)$$

where $a$ is as described in Problem 2.12.2.

On the other hand, if one uses the property of the logarithmic norm, $\mu(A(\omega) + B(t, \omega)) \le \mu(A(\omega)) + \mu(B(t, \omega))$, then in view of Theorem 2.9.1, the stability condition for (2.12.20) is given by

$$P\left\{\omega:\mu(A(\omega)) + \limsup_{t\to\infty}\left[\frac{1}{t-t_0}\int_{t_0}^{t}\mu(B(s, \omega))\,ds\right]\le -a\right\} = 1. \quad (2.12.22)$$

Furthermore, if we utilize the property that $|\mu(B(t, \omega))| \le \|B(t, \omega)\|$, then again in view of Theorem 2.9.1, the stability condition for (2.12.20) reduces to

$$P\left\{\Omega:\mu(A(\omega)) + \limsup_{t\to\infty}\left[\frac{1}{t-t_0}\int_{t_0}^{t}\|B(s, \omega)\|\,ds\right]\le -a\right\} = 1. \quad (2.12.23)$$

From the above development of stability conditions (2.12.21)–(2.12.23), one can easily see that condition (2.12.23) demands more than condition (2.12.22), and condition (2.12.22) demands more than condition (2.12.21).

Furthermore, if we assume that the elements $b_{ij}(t, \omega)$ of the random matrix function $B(t, \omega)$ are strictly stationary metrically transitive stochastic processes, then, by Theorem 2.1.5, stability conditions (2.12.21)–(2.12.23) reduce to

$$P\{\omega:\mu(A(\omega)) + E[B(t_0, \omega)] \le -a\} = 1, \quad (2.12.24)$$

$$P\{\omega:\mu(A(\omega)) + E[\mu(B(t_0, \omega))] \le -a\} = 1, \quad (2.12.25)$$

and

$$P\{\omega:\mu(A(\omega)) + E\|B(t_0, \omega)\| \le -a\} = 1, \quad (2.12.26)$$

respectively.

**Remark 2.12.3.**   By Problem 2.9.1, stability conditions (2.12.24)–(2.12.26) with respect to the norms (MN$_1$)–(MN$_5$) can easily be formulated. Nonetheless, in order to show the richness of the preceding stability analysis, we exhibit stability condition (2.12.25), corresponding to (MN$_3$); that is,

$$P\left\{\omega: -(a_{kk}(\omega) + E[b_{kk}(t_0,\omega)]) - \left(\sum_{\substack{i=1 \\ i \neq k}}^{n} |a_{ik}(\omega)| + E[|b_{ik}(t_0,\omega)|]\right) \geq a\right\} = 1,$$

(2.12.27)

which implies that at least one of the matrices $A(\omega)$ and $E[B(t_0,\omega)]$ is $P$-stable.

We shall next offer a simple example to illustrate the fact that the stability conditions presented can be interpreted in the context of the laws of large numbers.

Consider the linear differential system with random parameters

$$x'(t,\omega) = A(t,\omega)x(t,\omega), \qquad x(t_0,\omega) = x_0(\omega), \tag{2.12.28}$$

where $x \in R^n$, $E[\|A(t,\omega)\|]$ exists for $t \in R_+$, and the elements $a_{ij}(t,\omega)$ of the random matrix $A(t,\omega) = (a_{ij}(t,\omega))$ are product-measurable. Again, under these conditions the initial-value problem (2.12.28) has a unique sample solution for $t \geq t_0$. Set

$$C(t) = E[A(t,\omega)] \qquad \text{and} \qquad B(t,\omega) = A(t,\omega) - E[A(t,\omega)].$$

Then the system (2.12.28) can be rewritten as

$$x'(t,\omega) = C(t)x + B(t,\omega)x, \qquad x(t_0,x_0) = x_0(\omega). \tag{2.12.29}$$

Now analogously stability conditions (2.12.21)–(2.12.23) can be rewritten relative to (2.12.29). For example, conditions (2.12.22) and (2.12.23) become

$$P\left\{\omega: \limsup_{t\to\infty}\left[\frac{1}{t-t_0}\int_{t_0}^{t}\mu(C(s))\,ds\right] + \limsup_{t\to\infty}\left[\frac{1}{t-t_0}\int_{t_0}^{t}\mu(B(s,\omega))\,ds\right]\right.$$
$$\left. \leq -a\right\} = 1$$

(2.12.30)

and

$$P\left\{\omega: \limsup_{t\to\infty}\left[\frac{1}{t-t_0}\int_{t_0}^{t}\mu(C(s))\,ds\right] + \limsup_{t\to\infty}\left[\frac{1}{t-t_0}\int_{t_0}^{t}\|B(s,\omega)\|\,ds\right]\right.$$
$$\left. \leq -a\right\} = 1,$$

(2.12.31)

respectively. We note that conditions (2.12.30) and (2.12.31) show that the random function $A(t,\omega)$ satisfies the strong law of large numbers. Similarly, other laws of large numbers can be associated with other stability conditions such as (2.11.22).

**Remark 2.12.4.** From the foregoing discussion, we note that in the case of non-white-noise coefficients such as strictly stationary metrically transitive random coefficients, the randomness may be a stabilizing agent. This remark can be justified from (2.12.25) and (2.12.30).

To exhibit the fruitfulness of using a vector Lyapunov function instead a single Lyapunov function, we discuss the following example.

**Example 2.12.2.** Consider the system of differential equations with random coefficients

$$x'(t, \omega) = A(t, x, \omega)x(t, \omega), \qquad x(t_0, \omega) = x_0(\omega), \qquad (2.12.32)$$

where $x \in R^n$, $A(t, x, \omega)$ is a $2 \times 2$ random matrix function defined by

$$A(t, x, \omega) = \begin{bmatrix} e^{-t} - f_1(t, x, \omega) & \sin(2\pi t + \theta(\omega)) \\ \sin(2\pi t + \theta(\omega)) & e^{-t} - f_1(t, x, \omega) \end{bmatrix},$$

$f_1 \in M[R_+ \times R^2, R[\Omega, R_+]]$, $f_1(t, 0, \omega) \equiv 0$ w.p. 1, and $\theta \in R[\Omega, R_+]$ and is uniformly distributed over $[0, 2\pi]$. It is obvious that $\sin(2\pi t + \theta(\omega))$ is an ergodic process.

First, we choose a single Lyapunov function $V(t, x, \omega)$ that is given by $V(t, x, \omega) = x_1^2 + x_2^2$. It is easy to see that

$$D_{(2.12.32)}^+ V(t, x, \omega) \le [2e^{-t} + 2|\sin(2\pi t + \theta(\omega))|]V(t, x, \omega)$$

because $2|ab| \le a^2 + b^2$ and $f_1(t, x, \omega) \ge 0$. Clearly, the trivial solution process of the comparison equation

$$u'(t, \omega) = 2[e^{-t} + |\sin(2\pi t + \theta(\omega))|]u(t, \omega)$$

is not stable w.p. 1. Hence we cannot deduce any information about the stability of (2.12.32) from the scalar version of Theorem 2.11.1, even though it is stable.

Now we attempt to seek the stability analysis of (2.12.32) by employing the vector Lyapunov functions. We take

$$V(t, x, \omega) = \begin{bmatrix} (x_1 + x_2)^2 \\ (x_1 - x_2)^2 \end{bmatrix}.$$

It is easy to assume that components $V_1$ and $V_2$ of $V$ are not positive definite and hence do not satisfy the scalar version of Theorem 2.11.1. However, $V(t, x, \omega)$ does satisfy the hypotheses of Theorem 2.11.1. In fact,

$$(x_1^2 + x_2^2) \le \sum_{i=1}^{n} V_i(t, x, \omega) \le 2(x_1^2 + x_2^2),$$

and the vectorial inequality

$$D_{(2.12.32)}^+ V(t, x, \omega) \le g(t, V(t, x, \omega), \omega)$$

is satisfied with

$$g(t, u, \omega) = \begin{bmatrix} 2(e^{-t} + \sin(2\pi t + \theta(\omega))) & 0 \\ 0 & 2(e^{-t} - \sin(2\pi t + \theta(\omega))) \end{bmatrix} \begin{bmatrix} u_1 \\ u_2 \end{bmatrix}.$$

It is clear that $g(t, u, \omega)$ is sample quasi-monotone nondecreasing in $u$ for $t \in R_+$. It is obvious that $u'(t, \omega) = g(t, u(t, \omega), \omega)$ is stable w.p. 1. Consequently, the trivial solution of (2.12.32) is stable by Theorem 2.11.1.

**Problem 2.12.3.** Assume that the hypotheses of Theorem 2.11.5 hold. Then

(i)   $(SS_1^*)$ of (2.8.11) implies $(SS_1)$;
(ii)  $(SS_3^*)$ of (2.8.11) implies $(SS_3)$.

In the following, the method of variation of parameters will be used to derive the results on the stability w.p. 1 with respect to the trivial solutions of (2.8.1) and its nonlinear perturbation (2.7.9).

**Theorem 2.12.2.** Let the hypotheses of Theorem 2.11.6 hold. Then for $t_0 \in R_+$, there exists a positive number $a = a(t_0)$ such that

$$P\left\{\omega : \limsup_{t \to \infty} \left[ \frac{1}{t - t_0} \int_{t_0}^t \alpha(s, \omega)\, ds \right] \le -a(t_0) \right\} = 1 \qquad (2.12.33)$$

implies $(SS_3)$ of (2.8.1).

*Proof.* By following the proof of Theorem 2.11.6, one can arrive at (2.11.30). Now the proof of the theorem follows from relation (2.12.33).

**Theorem 2.12.3.** Let the hypotheses of Theorem 2.11.7 hold. Then relation (2.12.33) implies $(SS_3)$ of (2.7.9).

*Proof.* Let $\varepsilon > 0$ and be sufficiently small. Then relation (2.12.33) implies

$$P\left\{\omega : \lim_{t \to \infty} \left[ \varepsilon(t - t_0) + \int_{t_0}^t \alpha(s, \omega)\, ds \right] \le -a/2 \right\} = 1. \qquad (2.12.34)$$

Hence there exists a positive deterministic number $K$ such that

$$P\left\{\omega : \exp\left[ \varepsilon(t - t_0) + \int_{t_0}^t \alpha(s, \omega)\, ds \right] \le K, \, t \ge t_0 \right\} = 1. \qquad (2.12.35)$$

Because of hypothesis (iv), by using the argument in the proof of Theorem 2.11.7 with $\|x_0(\omega)\| \le \delta/K$, we arrive at (2.11.34). From (2.11.34) one can show that $\|y(t, t_0, x_0(\omega), \omega)\| < \delta$ for all $t \ge t_0$. Otherwise there exists a $T$ such that $\|y(t, t_0, x_0(\omega), \omega)\| = \delta$ and $\|y(t, t_0, x_0(\omega), \omega)\| \le \delta$ for $t_0 \le t \le T$ w.p. 1. Then from (2.11.34) and (2.12.35), we get the contradiction $\delta < (\delta/K)K = \delta$, which proves our claim. Thus (2.11.34) holds for all $t \ge t_0$ and this, together with (2.12.34), yields the desired conclusion.

## 2.13. STABILITY IN THE $p$th MEAN

Now, we shall present some stability criteria that assure the $p$th mean or moment stability properties of the trivial solutions of (2.8.1) and (2.7.9).

**Theorem 2.13.1.** Assume that conditions (i) and (ii) of Theorem 2.11.1 hold. Suppose further that

(iii)   for $(t, x) \in R_+ \times B(\rho)$

$$b(\|x\|^p) \leq \sum_{i=1}^{m} V_k(t, x, \omega) \leq a(t, \|x\|^p),$$

where $b \in \mathscr{VK}$, $a \in \mathscr{CK}$, and $p \geq 1$.

Then

$$(\text{SM}_1^*) \qquad \text{implies} \quad (\text{SM}_1).$$

*Proof.*   Let $0 < \varepsilon < \rho$, $t_0 \in R_+$ be given. Assume that (SM$_1^*$) holds. Then for $b(\varepsilon) > 0$ and $t_0 \in R_+$, there exists $\delta_1 = \delta(t_0, \varepsilon)$ such that

$$\sum_{i=1}^{m} E[u_{i0}(\omega)] \leq \delta$$

implies

$$\sum_{i=1}^{m} E[u_i(t, t_0, u_0, \omega)] < b(\varepsilon^p), \qquad t \geq t_0. \tag{2.13.1}$$

We chose $u_0(\omega)$ such that $V(t_0, x_0(\omega), \omega) \leq u_0(\omega)$ and

$$\sum_{i=1}^{m} E[u_{i0}(\omega)] = a(t_0, E[\|x_0(\omega)\|^p]) \qquad \text{for} \quad x_0 \in B(\rho). \tag{2.13.2}$$

Since $a(t_0, \cdot) \in \mathscr{K}$, we can find a $\delta = \delta(t_0, \varepsilon)$ such that

$$E[\|x_0\|^p] \leq \delta \qquad \text{implies} \qquad a(t_0, E[\|x_0(\omega)\|^p]) \leq \delta_1. \tag{2.13.3}$$

Now we claim that if $(E[\|x_0\|^p])^{1/p} \leq \delta$, then $(E[\|x(t)\|^p])^{1/p} < \varepsilon$, $t \geq t_0$. Suppose that this is false. Then there would exist a solution process $x(t, t_0, x_0, \omega)$ with $(E[\|x\|^p])^{1/p} \leq \delta$ and a $t_1 \geq t_0$ such that

$$(E[\|x(t_1)\|^p])^{1/p} = \varepsilon \quad \text{and} \quad (E[\|x(t)\|^p])^{1/p} \leq \varepsilon, \qquad t \in [t_0, t_1]. \tag{2.13.4}$$

By following the proof of Theorem 2.11.1 and using he convexity of $b$, we have

$$b(\varepsilon^p) \leq \sum_{i=1}^{m} E[V_i(t_1, x(t_1, \omega), \omega)] \leq \sum_{i=1}^{m} E[r_i(t_1, t_0, u_0, \omega)]. \tag{2.13.5}$$

Relations (2.13.1), (2.13.4), and (2.13.5) lead to the contradiction

$$b(\varepsilon^p) \leq \sum_{i=1}^{m} E[V_i(t_1, x(t_1, \omega), \omega)] \leq \sum_{i=1}^{m} E[r_i(t_1, t_0, u_0, \omega)] < b(\varepsilon^p),$$

proving (SM$_1$).

**Example 2.13.1.**   Let $\mu(A(t, \omega))$ be the logarithmic norm of $A(t, \omega)$ in (2.9.15). Then for some Lebesgue-integrable function $\lambda$ on $[t_0, \infty)$ defined on $R_+$ into $R$,

$$E\left[\exp\left[p \int_{t_0}^{t} \mu(A(s, \omega)) \, ds\right]\right] \leq \exp\left[\int_{t_0}^{t} \lambda(s) \, ds\right] \tag{2.13.6}$$

implies (SM$_1$) of (2.9.15).

*Proof.*   Take $V(t, x, \omega) = \|x\|^p$, $p \geq 1$, and obtain

$$D^+_{(2.9.15)} V(t, x, \omega) \leq p\mu(A(t, \omega)) V(t, x, \omega).$$

From (2.13.6) it is easy to see that the trivial solution of

$$u'(t, \omega) = p\mu(A(t, \omega)) u(t, \omega), \qquad u(t_0, \omega) = u_0(\omega),$$

is stable in the mean whenever $u_0(\omega)$ and $\mu(A(t, \omega))$ are independent random variables or $u_0(\omega)$ is a.s. bounded. Hence by the application of Theorem 2.13.1, we can conclude that the trivial solution of (2.9.15) satisfies (SM$_1$).

As before in Sections 2.11 and 2.12, it is now easy to formulate and prove the notions corresponding to (SM$_2$)–(SM$_4$) based on the previous proofs. We shall again incorporate these results in the problem.

**Problem 2.13.1.**   Let the hypotheses of Theorem 2.13.1 be satisfied. Then show that (SM$_3^*$) implies (SM$_3$). Further, if $a(t, u) \equiv a(u)$, then show that (SM$_2^*$) implies (SM$_2$) and (SM$_4^*$) implies (SM$_4$).

**Remark 2.13.1.**   In Theorem 2.13.1, we have assumed that the trivial solution of comparison system (2.10.1) has the first-mean stability property and $V(t, x, \omega)$ satisfies the estimate

$$b(\|x\|^p) \leq \sum_{i=1}^{m} V_i(t, x, \omega) \leq a(t, \|x\|^p).$$

These conditions can be replaced by the assumptions, namely,

$$b(\|x\|^p) \leq \left[\sum_{i=1}^{m} V_i(t, x, \omega)\right]^p \leq a(t, \|x\|^p),$$

and the trivial solution of the comparison system (2.10.1) has the $p$th mean stability property.

**Example 2.13.2.**   Let us consider the scalar random differential equation

$$x'(t, \omega) = (a + \eta(t, \omega))x(t, \omega), \qquad x(t_0, \omega) = x_0(\omega), \qquad (2.13.7)$$

where $x \in R$, $\eta(t, \omega)$ is a stationary Gaussian process with mean $E[\eta(t, \omega)] = 0$, and the covariant function $C(t - s) = E[\eta(t, \omega)\eta(s, \omega)]$.

From the properties of Gaussian process, it is obvious that relation (2.13.6) becomes

$$E\left[\exp\left[p \int_{t_0}^t (a + \eta(s, \omega)) \, ds\right]\right] = \exp\left[pa(t - t_0) + \frac{p^2}{2} \int_{t_0}^t \int_{t_0}^t C(u - s) \, du \, ds\right].$$

The trivial solution of (2.13.7) is uniformly stable in $p$th mean if

$$\left|a(t - t_0) + \frac{p}{2} \int_{t_0}^t \int_{t_0}^t C(u - s) \, du \, ds\right| < \infty.$$

This implies that the process $\eta(t, \omega)$ has bounded spectral density $d(\lambda)$ and

$$\int_{t_0}^t \int_{t_0}^t C(u - s) \, du \, ds = d(0)(t - t_0) + O(t - t_0). \qquad (2.13.8)$$

In fact, the trivial solution of (2.13.7) is uniformly asymptotically stable in the $p$th mean whenever

$$2a + pd(0) < 0.$$

Finally, by using the method of variation of parameters, we give some sufficient conditions for the $p$th moment stability of the trivial solution of (2.8.1) and its random perturbation (2.7.9).

**Theorem 2.13.2.**   Assume that the hypotheses of Theorem 2.11.6 hold. Further, assume that there exists a deterministic Lebesgue-integrable function $\lambda$ on $R_+$ defined on $R_+$ into $R$ and a positive number $a = a(t_0)$ such that

$$E\left[\exp\left[p \int_{t_0}^t \alpha(s, \omega) \, ds\right]\right] \leq \exp\left[\int_{t_0}^t \lambda(s) \, ds\right] \qquad (2.13.9)$$

and

$$-a(t_0) \equiv \limsup_{t \to \infty} \left[\frac{1}{t - t_0} \int_{t_0}^t \lambda(s) \, ds\right]. \qquad (2.13.10)$$

Then the trivial solution of (2.8.1) is $p$th mean asymptotically stable (SM₃).

*Proof.*   By imitating the proof of Theorem 2.11.6, we obtain

$$\|x(t, t_0, x_0(\omega), \omega)\| \leq \|x_0(\omega)\| \exp\left[\int_{t_0}^t \alpha(s, \omega) \, ds\right], \qquad t \geq t_0.$$

First, taking the $p$th exponent and then the expected value of the resulting inequality and using the fact that $x_0(\omega)$ is independent of $\alpha(t, \omega)$, we get

$$E[\|x(t, t_0, x_0(\omega), \omega)\|^p] \leq E[\|x_0(\omega)\|^p]E\left[\exp\left[p \int_{t_0}^t \alpha(s, \omega)\, ds\right]\right] \quad (2.13.11)$$

for $t \geq t_0$. From (2.13.10), it is obvious that

$$\lim_{t \to \infty}\left[\exp\left[\int_{t_0}^t \lambda(s)\, ds\right]\right] = 0. \quad (2.13.12)$$

From (2.13.9), (2.13.11), and (2.13.12), we conclude that the trivial solution of (2.8.1) is $p$th mean asymptotically stable.

**Theorem 2.13.3.** Let the hypotheses of Theorem 2.11.7 hold. Then relations (2.13.9) and (2.13.10) imply (SM$_3$) of (2.7.9).

*Proof.* The proof of this theorem can be formulated based on the proofs of Theorems 2.11.7 and 2.13.2. The details are left as an exercise.


**Notes**

The introductory discussion on sample approach is adapted from the standard books on probability theory and the theory of deterministic and random functions, namely, Bartlett [3], Doob [14], Gikhman and Skorokhod [19], Natanson [69], and Wong [87]. The existence theorem (Theorem 2.1.1) is new and is based on Itô and Nisio [34]. See also, Morozan [68], Edsinger [16], Soong [81], and Strand [84]. The contents of Section 2.3 are new. They are natural extensions of deterministic results given in Lakshmikantham and Leela [59] regarding differential inequalities in the sense of Carathéodory type. See also Ladde [51]. The results on maximal and minimal solutions are new. Theorem 2.5.1 is taken from the work of Ladde [51]. See also Khas'minskii [38]. The uniqueness and continuous dependence theorems are new and are based on deterministic results of Lakshmikantham and Leela [59] and stochastic results of Morozan [68], Edsinger [16], Soong [81], and Strand [84]. Most of the material of Section 2.7 is developed from the deterministic results of Lakshmikantham and Leela [59] and Lord and Mitchell [62]. Theorems 2.8.1 and 2.8.2 are taken from Ladde [51], whereas Theorem 2.8.3 is new. The results of Section 2.9 are due to Ladde [52, 53]. The contents of Sections 2.11, 2.12 and 2.13 are adapted from Ladde [51, 52], Khas'minskii [39], and Morozan [66–68]. For related results, see Bertram and Sarachik [4], Bunke [7–10], Caughey [11], Caughey and Gray [12], Kats [35], Kats and Krasovskii [36], Khas'minskii [37, 40, 41], Kozin [43, 44, 46], Ladde, Lakshmikantham, and Liu [55], Leibowitz [60], Samuels [77, 78], and Shur [79].

# CHAPTER 3

# $L^p$-CALCULUS APPROACH

### 3.0. INTRODUCTION

This chapter is devoted to the study of random differential equations in the framework of $p$th moment or $L^p$-calculus. In Section 3.1 we briefly discuss $L^p$-calculus. We begin with $L^p$-continuity and differentiability of random processes. Next we discuss the $L^p$-integral (Bochner integral) and present some results about the sample Lebesgue and $L^p$-integrability of a random process. Section 3.2 concerns the relationship between sample and $L^p$-solutions of (2.0.1). We begin Section 3.3 by giving a simple example which shows the limitations of a Lipschitz condition for establishing the existence and uniqueness of the $L^p$-solutions of a random differential system. We then prove an existence and uniqueness result for $L^p$-solutions of (2.0.1), assuming a sample path Lipschitz condition. Moreover, we consider a Peano-type existence result that gives sufficient conditions under which every sample solution of (2.0.1) is also an $L^p$-solution. Several examples are given in order to demonstrate the applicability of the results. Furthermore, the conditions obtained are explicit and relate the statistics of the rate functions. In Section 3.4 the continuous dependence of $L^p$-solutions with respect to parameters is discussed. Also, results concerning $L^p$-continuity and $L^p$-differentiability of solutions with respect to initial data are developed which can be used to prove variation of constants formulas in the present framework. By using Lyapunov functions and the theory of deterministic differential inequalities, very general comparison theorems, which can be utilized to study stability theory, are obtained in Section 3.5. Finally, in Section 3.6 some stability results are given in the framework of Lyapunov functions and the theory of differential inequalties.

### 3.1. $L^p$-CALCULUS

This section deals with some basic concepts and results regarding $p$th moment or mean calculus. Let $L^p[\Omega, R^n]$ be a Banach space of equivalence classes of random functions defined on a complete probability space $(\Omega, \mathscr{F}, P)$ into $R^n$ with finite $p$th moment $(1 \leq p < \infty)$. We let $\|x\|_p = (\int_\Omega \|x(\omega)\|^p P(d\omega))^{1/p}$. Let us denote by $R[[a, b], L^p[\Omega, R^n]]$ the collection of functions defined on $[a, b]$ with values in $L^p[\Omega, R^n]$.

**Definition 3.1.1.** A function $x \in R[[a, b], L^p[\Omega, R^n]]$ is said to be *continuous in the pth moment or in the mean* ($L^p$-*continuous or strongly continuous*) at $t \in (a, b)$ if

$$(E \lfloor \|x(t + h) - x(t)\|^p])^{1/p} \to 0 \qquad \text{as} \quad h \to 0.$$

If $x$ is $L^p$-continuous for each $t \in (a, b)$ and if it is one-sided continuous at the end points $a$ and $b$, then $x$ is said to be $L^p$-continuous on $[a, b]$.

Let $C[[a, b], L^p[\Omega, R^n]]$ be a collection of $L^p$-continuous functions defined on $[a, b]$ with values in $L^p[\Omega, R^n]$.

**Definition 3.1.2.** A function $x \in C[[a, b], L^p[\Omega, R^n]]$ is said to have a *pth moment or mean derivative* ($L^p$-*derivative or strong derivative*) $x'(t)$ at $t \in (a, b)$ if

$$\left( E \left[ \left\| \frac{x(t + h) - x(t)}{h} - x'(t) \right\|^p \right] \right)^{1/p} \to 0 \qquad \text{as} \quad h \to 0.$$

If $x$ has a $p$th moment derivative for each $t \in (a, b)$ and possesses a one-sided derivative at the end points $a$ and $b$, then $x$ is said to be $L^p$-differentiable on $[a, b]$.

**Definition 3.1.3.** A function $x \in R[[a, b], L^p[\Omega, R^n]]$ is said to have a $p$th moment or mean integral (or $L^p$-integral) on $[a, b]$ if $x(t)$ has a Bochner integral on $[a, b]$.

We note that the $L^p$-continuity, differentiability, and integrability obey all elementary properties of the calculus of abstract functions defined on $[a, b]$ with values in a Banach space. For details, see Dunford and Schwartz [15] and Ladas and Lakshmikantham [49]. However, for easy reference, we list some important results that will be needed in our subsequent discussion.

The following result gives a necessary and sufficient condition for $L^p$-integrability.

**Lemma 3.1.1.** A function $x \in R[[a,b], L^p[\Omega, R^n]]$ is $L^p$-integrable on $[a,b]$ if and only if it is $L^p$-measurable ($\|x(t)\|_p$ is a measurable function on $[a,b]$) and $\|x(t)\|_p$ is Lebesgue-integrable on $[a,b]$.

The next result establishes the relationship between the indefinite $L^p$-integral and $L^p$-absolute continuity.

**Lemma 3.1.2.** A function $x(t)$ is $L^p$-absolutely continuous and its $L^p$-derivative $x'(t) \in L^p[\Omega, R^n]$ for almost every $t \in [a,b]$ if and only if

$$x(t) = x(a) + \int_a^t x'(s)\,ds,$$

where the integral is in the $L^p$-sense.

Suppose that $([a,b], \mathscr{F}^1, m)$ is a Lebesgue-measure space and $F$ is a Lebesgue-measurable function on $[a,b]$ whose values are in $L^p[\Omega, R^n]$, $1 \le p < \infty$. For each $t \in [a,b]$, $F(t) = [f(t,\omega)]$ is an equivalence class of random functions $f(t,\omega)$, defined on a complete probability space $(\Omega, \mathscr{F}, P)$ into $R^n$. If for each $t \in [a,b]$ we select a particular random variable $f(t,\omega) \in F(t)$, the resulting function $f(t,\omega)$, defined on $([a,b], \mathscr{F}^1, m) \times (\Omega, \mathscr{F}, P) \equiv [a,b] \times \Omega$, will be called a representation of the function $F$. The results that follow establish the relation between the product measurability of representation of $F$ and $L^p$-integrability of. $F$.

**Lemma 3.1.3.** If $F \in R[[a,b], L^1[\Omega, R^n]]$ and if $F(t)$ is first-moment-integrable on $[a,b]$, then there exists a product-measurable random function $f$ on $[a,b] \times \Omega$ into $R^n$ such that $[f(t,\omega)] = F(t)$ for almost all $t \in [a,b]$. Moreover, its sample Lebesgue integral

$$\int_a^b f(t,\omega)\,dt$$

exists w.p. 1 and

$$\left[ \int_a^b f(t,\omega)\,dt \right] = \int_a^b F(t)\,dt,$$

where the right-hand integral is in the sense of $L^p$ with $p = 1$.

**Lemma 3.1.4.** Let $1 \le p < \infty$, and let $f \in M[[a,b], R[\Omega, R^n]]$ such that $F(t) = [f(t,\omega)] \in L^p[\Omega, R^n]$ for almost all $t \in [a,b]$. Then the function $F$ defined on $[a,b]$ into $L^p[\Omega, R^n]$ is $L^p$-measurable on $[a,b]$.

The following result shows that the $L^p$-integrability of a function implies its sample integrability.

**Theorem 3.1.1.** Let $1 \le p < \infty$, and let $F$ be a $p$th mean integrable function on $[a,b]$ into $L^p[\Omega, R^n]$. Then there exists a product-measurable function defined on $[a, b] \times \Omega$ into $R^n$ which is uniquely determined except for a set whose product measure is zero and such that $F(t) = [f(t, \omega)]$ for almost all $t \in [a, b]$. Moreover, $f(t, \omega)$ is sample Lebesgue-integrable on $[a, b]$ w.p. 1, and the sample Lebesgue integral

$$\int_a^b f(t, \omega)\, dt \quad \text{w.p. 1}$$

belongs to $R[\Omega, R^n]$, and its equivalence class satisfies

$$\left[ \int_a^b f(t, \omega)\, dt \right] = \int_a^b F(t)\, dt,$$

where the right-hand integral is in the sense of the $L^p$-integral.

A result that permits the passage to the limit of a $L^p$-integrable sequence under the integral sign is given in

**Theorem 3.1.2.** (generalized Lebesgue dominated convergence theorem). Let $1 \le p < \infty$, and $([a, b], \mathscr{F}^1, m)$ be a Lebesgue-measurable space. For $n = 1, 2, \ldots, u_n, u \in R[[a, b], L^p[\Omega, R^n]]$, $\{u_n\}$ is a sequence of $L^p$-integrable functions which converges a.e. on $[a, b]$ to an $L^p$-integrable function $u$. Let $x_n$ be a sequence of measurable functions such that $|x_n(t)| \le u_n(t)$ and $x_n(t)$ converges to $x(t)$ w.p. 1 and a.e. on $[a, b]$. If

$$\lim_{n \to \infty} \int_a^b u_n(t)\, dt = \int_a^b u(t)\, dt,$$

then

$$\lim_{n \to \infty} \int_a^b x_n(t)\, dt = \int_a^b x(t)\, dt,$$

where the convergence and integral are in the $L^p$-sense.

We note that the above results are simplified versions of more general results that are available in the literature.

## 3.2. INTERRELATIONSHIP BETWEEN SAMPLE AND $L^p$-SOLUTIONS

Let us recall that $L^p[\Omega, R^n]$ is a Banach space of equivalence classes of random functions defined on a complete probability space $(\Omega, \mathscr{F}, P)$ into $R^n$ with finite $p$th moment $(1 \le p < \infty)$. We employ the following notations:

$[R_+ \times L^p[\Omega, R^n], [L^p[\Omega, R^n]]]$ is the set of all functions $f(t, x)$ defined on $R_+ \times L^p[\Omega, R^n]$ into $L^p[\Omega, R^n]$. For fixed $z \in L^p[\Omega, R^n]$ and a positive number $\rho$, we let $B(z, \rho) = \{x \in L^p[\Omega, R^n] : \|x - z\|_p < \rho\}$ and $\bar{B}(z, \rho)$ be the closure of $B(z, \rho)$.

Let us consider a system of first-order differential equations

$$x' = f(t, x), \qquad x(t_0) = x_0, \tag{3.2.1}$$

where the prime denotes the $p$th moment derivative of $x$ and

$$f \in [J \times L^p[\Omega, R^n], L^p[\Omega, R^n]],$$

$J$ being, as before, the interval $[t_0, t_0 + a]$, $t_0 \geq 0$.

**Definition 3.2.1.** A function $x : J \to L^p[\Omega, R^n]$ is said to be a solution of the initial-value problem (3.2.1) if it satisfies the following conditions:

(i)   $x(t_0) = x_0$,
(ii)  $x(t)$ is $p$th mean continuous, and
(iii) $x(t)$ is $p$th mean differentiable for almost every $t \in J$ and satisfies (3.2.1).

It is clear that by (ii) and (iii) $x(t)$ is $p$th mean absolutely continuous.

Assume that $f$ is a function defined on $J \times L^p[\Omega, R^n]$ into $L^p[\Omega, R^n]$ and $f(t, x(t))$ is Bochner-integrable on $J$ whenever $x(t)$ is strongly continuous on $J$. Then we observe by Theorem 3.1.1 that the initial-value problem (3.2.1) is equivalent to the integral equation

$$x(t) = x_0 + \int_{t_0}^{t} f(s, x(s)) \, ds, \tag{3.2.2}$$

where the integral represents the $L^p$-integral or the Bochner integral.

We now show that the sample problem (2.2.1), namely,

$$x'(t, \omega) = f(t, x(t, \omega), \omega), \qquad x(t_0, \omega) = x_0(\omega), \tag{3.2.3}$$

can be formulated as an $L^p$-problem (3.2.1). To do this, we simply set $f(t, x) = [f(t, x(t, \omega), \omega)]$ and consider $f(t, x)$ an element of $L^p(\Omega, R^n)$.

The following result establishes the relationship between the sample and $L^p$ initial-value problems.

**Theorem 3.2.1.** (1)  If $x(t)$ is an $L^p$-solution of (3.2.1), then there exists a sample solution $x(t, \omega)$ of (3.2.3) such that $x(t) = [x(t, \omega)]$.

(2)  If $x(t, \omega)$ is a sample solution of (3.2.3), then the equivalent class $x(t) = [x(t, \omega)]$ is the $L^p$-solution of (3.2.1) if and only if

$$\int_{t_0}^{t} \| f(s, x(s)) \|_p \, ds < \infty \qquad \text{for} \quad t \in J \tag{3.2.4}$$

and $x_0 \in L^p[\Omega, R^n]$.

*Proof.* To prove (1), first notice that any $L^p$-solution $x(t)$ satisfies the integral equation (3.2.2). Therefore, by Theorem 3.1.1 we conclude that

$$x_0 + \int_{t_0}^t f(s, x(s))\, ds = [x_0(\omega)] + \left[ \int_{t_0}^t f(s, x(s, \omega), \omega)\, ds \right],$$

and hence

$$x(t, \omega) = x_0(\omega) + \int_{t_0}^t f(s, x(s, \omega), \omega)\, ds \quad \text{w.p. } 1,$$

which implies by Lemma 2.2.1 that $x(t, \omega)$ is a sample solution of (3.2.3). This completes the proof of (1).

To prove (2), we first prove that a sample solution $x(t, \omega)$ that satisfies (3.2.4) and $x_0 \in L^p[\Omega, R^n]$ is an $L^p$-solution. We note that $f(t, x(t, \omega), \omega)$ is product-measurable and $[f(t, x(t, \omega), \omega)] = f(t, x(t)) \in L^p[\Omega, R^n]$ a.e. on $J$. Then by Lemma 3.1.4 the function $f(t, x(t))$ is $L^p$-measurable on $J$. This, together with (3.2.4), implies that $f(t, x(t))$ is $L^p$-integrable. Hence by Theorem 3.1.1 we conclude that

$$\int_{t_0}^t f(s, x(s))\, ds = \left[ \int_{t_0}^t f(s, x(s, \omega), \omega)\, ds \right],$$

and thus

$$x_0 + \int_{t_0}^t f(s, x(s))\, ds = [x_0(\omega)] + \left[ \int_{t_0}^t f(s, x(s, \omega), \omega)\, ds \right].$$

Therefore, $x(t) = [x(t, \omega)]$ is an $L^p$-solution of (3.2.1). To complete the proof of (2), we need to show that if $x(t) = [x(t, \omega)]$ is an $L^p$-solution, then (3.2.4) holds. The proof of the statement follows by the application of Lemma 3.1.1.

In the light of the foregoing discussion, it is easy to see that several results presented in Chapter 2 (for example, results in Sections 2.3–2.5, 2.8, and 2.9) can be interpreted in the context of equivalence classes of random functions. We leave the details to the reader.

## 3.3. EXISTENCE AND UNIQUENESS

It is well-known that the basic existence theorem of Peano- or Carathéodory-type for (3.2.1) fails unless $f(t, x)$ satisfies additional conditions. The reason for this is that in infinite-dimensional Banach space, a closed unit ball is not necessarily compact. However, if $f(t, x)$ satisfies a Lipschitz condition in $x$, then the local existence and uniqueness of solutions of (3.2.1) is guaranteed; see [49].

The following example demonstrates the limited applicability of the Lipschitz condition.

**Example 3.3.1.** Consider the $L^p$ initial-value problem

$$x'(t) = Ax(t), \qquad x(0) = x_0, \tag{3.3.1}$$

where $A \in R[\Omega, R]$. If $f(t, x) = Ax$ is to satisfy a Lipschitz condition, then we need to assume that there exists a positive constant $K$ such that

$$\|Ax\|_p \leq K\|x\|_p \tag{3.3.2}$$

for all $x \in L^p[\Omega, R]$. However, $f(t, x)$ satisfies (3.3.2) if and only if $A(\omega)$ is a.s. bounded, that is, there exists an $M$ such that $|A(\omega)| \leq M$ for almost all $\omega \in \Omega$. The sufficiency of the boundedness condition is obvious with $K = M$. To prove necessity, we note that each $L^p$-solution is also a sample solution. But $x(t) = e^{tA}x_0$ is the only possible $L^p$-solution. We set $T = e^{tA}$ for some fixed $t \in J$ and consider the map $T : L^p[\Omega, R] \rightarrow L^p[\Omega, R]$, defined by $Tx_0 = e^{tA}x_0$. We note that $T$ is a closed linear map. By the application of the closed graph theorem, we conclude that $T$ is bounded. This proves that $|A(\omega)| \leq M$.

The foregoing example shows that one cannot expect linear differential equations whose coefficients have Gaussian or Poisson distributions to satisfy a Lipschitz condition in the $L^p$-sense.

We shall now prove a result on a sample path Lipschitz condition on $f$.

**Theorem 3.3.1.** Assume that

(i)   $f \in M[J \times \bar{B}(z, \rho), R[\Omega, R^n]]$ and $f(t, x, \omega)$ is sample continuous in $x$ for each $t \in J$;

(ii)   $x_0 \in L^p[\Omega, R^n]$ and $f(t, x_0(\omega), \omega)$ is $L^p$-integrable on $J$;

(iii)   there exists a product-measurable function $K \in M[J, R[\Omega, R_+]]$, which is sample Lebesgue-integrable on $J$ and satisfies the sample Lipschitz condition

$$\|f(t, x, \omega) - f(t, y, \omega)\| \leq K(t, \omega)\|x - y\| \tag{3.3.3}$$

w.p. 1 for all $x, y \in \bar{B}(z, \rho)$;

(iv)   the linear homogeneous random differential equation

$$u'(t, \omega) = K(t, \omega)u(t, \omega), \qquad u(t_0, \omega) = u_0(\omega) \tag{3.3.4}$$

with

$$u(t_0, \omega) = \int_{t_0}^{t_0 + a} \|f(s, x(t_0, \omega), \omega)\| \, ds \tag{3.3.5}$$

has a unique $L^p$-solution on $J$. Then the initial-value problem (3.2.3) has a unique $L^p$-solution on $J$.

*Proof.* By Corollary 2.6.1 and Remark 2.6.1, it follows that the initial-value problem (3.2.3) has a unique sample solution $x(t, \omega)$ on $J$. Our objective now is to prove that $x(t) = [x(t, \omega)]$ is the unique $L^p$-solution of (3.2.1).

By Theorem 3.2.1, it is enough to show that $f(t, x(t, \omega), \omega)$ satisfies relation (3.2.4).

First, we consider the following approximations relative to the initial-value problem (3.3.4). Setting $L(\omega) = \int_{t_0}^{t_0 + a} \|f(s, x_0(\omega), \omega)\|\, ds$, we see because of (ii) that $L \in L^p[\Omega, R_+]$. Define

$$u_0(t, \omega) = 0, \qquad u_{n+1}(t, \omega) = L(\omega) + \int_{t_0}^t K(s, \omega) u_n(s, \omega)\, ds \qquad \text{for} \quad t \in J,$$

where the integral is the sample Lebesgue integral. It is obvious that the above sequence satisfies the following properties:

(a)  for each $n$, $u_n(t, \omega)$ is a nondecreasing function of $t$;
(b)  $u_n(t, \omega)$ is a nondecreasing sequence;
(c)  $u_n(t, \omega) = (1 + \sum_{i=1}^n (1/i!)(\int_{t_0}^t K(s, \omega)\, ds)^i) L(\omega)$;
(d)  $u_n(t, \omega) \to u(t, \omega)$ uniformly in $t$ w.p. 1 uniformly in $L^p$ and in the $L^p$-integral sense, where $u(t, \omega)$ is a sample solution of (3.3.4);
(e)  $u_n'(t, \omega) \geq 0$ w.p. 1 and is a nondecreasing sequence;
(f)  $u_n'(t, \omega) \to u(t, \omega)$ w.p. 1 for almost all $t \in J$ in $L^p$ and in the $L^p$-integral sense.

Now define

$$x_0(t, \omega) - x_0, \qquad x_{n+1}(t, \omega) = x_0(\omega) + \int_{t_0}^t f(s, x_n(s, \omega), \omega)\, ds.$$

By Remark 2.6.1, it follows that $x_{n+1}(t, \omega) \to x(t, \omega)$ uniformly in $t$ w.p. 1, where $x(t, \omega)$ is a sample solution of (3.2.3). Note that

$$\|x_{n+1}(t, \omega) - x_n(t, \omega)\| \leq \int_{t_0}^t K(s, \omega) \|x_n(s, \omega) - x_{n-1}(s, \omega)\|\, ds.$$

Since

$$\|x_1(t, \omega) - x_0(\omega)\| \leq \int_{t_0}^t \|f(s, x_0(\omega), \omega)\|\, ds \leq L(\omega)$$

$$\leq u_1(t, \omega) - u_0(t, \omega)$$

and

$$u_{n+1}(t, \omega) - u_n(t, \omega) = \int_{t_0}^t K(s, \omega)(u_n(s, \omega) - u_{n-1}(s, \omega))\, ds,$$

it is easy to see that

$$\|x_{n+1}(t, \omega) - x_n(t, \omega)\| \leq u_{n+1}(t, \omega) - u_n(t, \omega)$$

and

$$\|x_{n+1}'(t, \omega) - x_n'(t, \omega)\| \leq K(t, \omega)(u_n(t, \omega) - u_{n-1}(t, \omega))$$

$$\leq u_{n+1}'(t, \omega) - u_n'(t, \omega).$$

Hence by (d), $x_n(t, \omega) \to x(t, \omega)$ uniformly in $t$ w.p. 1, uniformly in $L^p$, and in the $L^p$-integral sense. Similarly, by (f), $x_n'(t, \omega) \to x'(t, \omega)$ w.p. 1 for almost all $t$, in $L^p$ for almost all $t$, and in the $L^p$-integral sense.

We note that

$$f(t, x_n(t, \omega), \omega) = f(t, x_0(\omega), \omega)$$

$$+ \sum_{i=1}^{n} \left[ f(t, x_i(t, \omega), \omega) - f(t, x_{i-1}(t, \omega), \omega) \right]$$

and hence that

$$\| f(t, x_n(t, \omega), \omega) \| \leq \| f(t, x_0(\omega), \omega) \|$$

$$+ \sum_{i=1}^{n} K(t, \omega) \| x_i(t, \omega) - x_{i-1}(t, \omega) \|$$

$$\leq \| f(t, x_0(\omega), \omega) \| + \sum_{i=1}^{n} (u_{i+1}'(t, \omega) - u_n'(t, \omega))$$

$$\leq \| f(t, x_0(\omega), \omega) \| + u_{n+1}'(t, \omega) - u_0'(t, \omega)$$

$$\leq \| f(t, x_0(\omega), \omega) \| + u_{n+1}'(t, \omega)$$

$$\leq \| f(t, x_0(\omega), \omega) \| + u'(t, \omega).$$

Consequently,

$$\| f(t, x_n(t)) \|_p \leq \| f(t, x_0) \|_p + \| u'(t) \|_p.$$

Also,

$$\| f(t, x_n(t)) \|_p \to \| f(t, x(t)) \|_p$$

as $n \to \infty$ for almost all $t \in J$. Thus by the application of Lebesgue dominated convergence theorem we conclude that

$$\int_{t_0}^{t} \| f(t, x(t)) \|_p \, dt < \infty \qquad \text{on} \quad J.$$

The proof of uniqueness follows from the uniqueness of sample solutions of (3.2.3)

**Corollary 3.3.1.**   Assume that all hypotheses of Theorem 3.3.1 hold, except that (ii) and (iv) are replaced by

$$\int_{t_0}^{t} \left\| K(s, \omega) \int_{t_0}^{t} \| f(v, x_0(\omega), \omega) \| \, dv \exp \left[ \int_{t_0}^{s} K(u, \omega) \, du \right] \right\|_p \, ds < \infty$$

$$\text{on} \quad J. \quad (3.3.6)$$

Then the conclusion of Theorem 3.3.1 remains true.

*Proof.*  It is enough to show that the sample solution $u(t, \omega)$ is an $L^p$-solution of (3.3.4). This means that it is sufficient to show that

$$\int_{t_0}^t \|K(s, \omega)u(s, \omega)\|_p \, ds < \infty \qquad \text{for} \quad t \in J, \tag{3.3.7}$$

that is,

$$\int_{t_0}^t \left\| K(s, \omega)u_0(\omega) \exp\left[ \int_{t_0}^s K(u, \omega) \, du \right] \right\|_p \, ds < \infty \qquad \text{on} \quad J,$$

where $u_0(\omega)$ is as defined in (3.3.5). This implies that

$$\int_{t_0}^t \left\| K(s, \omega) \int_{t_0}^{t_0+a} \| f(v, x_0(\omega), \omega) \| \, dv \exp\left[ \int_{t_0}^s K(u, \omega) \, du \right] \right\|_p \, ds < \infty \qquad \text{on} \quad J.$$

Hence by (3.3.6), (3.3.7) is true, proving the corollary.

**Problem 3.3.1.**  Consider the linear sample problem

$$x'(t, \omega) = A(t, \omega)x(t, \omega) + P(t, \omega), \qquad x(t_0\omega) = x_0(\omega), \tag{3.3.8}$$

where $A(t, \omega)$ is an $n \times n$ product-measurable random matrix function whose entries $a_{ij}(t, \omega)$ are sample Lebesgue-integrable on $J$, $P(t, \omega)$ is a product-measurable $n$-dimensional random vector function whose elements are $L^r$-integrable on $J$, and $x_0 \in L^p[\Omega, R^n]$. Let $K(t, \omega) = \|A(t, \omega)\|$. Suppose that $A(t, \omega)$, $P(t, \omega)$, and $x_0(\omega)$ are independent random functions for all $t \in J$. Let $L(s, t) = E(\exp(sK(t, \omega)))$. Then show that the sample solution $x(t, \omega)$ of (3.3.8) is an $L^p$-solution (unique) on $J$ if

$$\int_{t_0}^t L(s^u, u) \, du < \infty \qquad \text{for} \quad t \in J, \tag{3.3.9}$$

and

$$p < s^* a^{-1}. \tag{3.3.10}$$

(*Hint:*  By using the Hölder inequality, Lemma 3.1.4, and the properties of a moment-generating function $L(s, t)$ of $K(t, \omega)$, show that

$$\int_{t_0}^t \left\| K(s, \omega) \int_{t_0}^{t_0+a} K(v, \omega) \, dv \exp\left( \int_{t_0}^s K(u, \omega) \, du \right) \right\|_p \, ds < \infty \qquad \text{on} \quad J.)$$

**Problem 3.3.2.**  Consider the sample initial-value problem (3.3.8) in which $A(t, \omega)$, $P(t, \omega)$, and $x_0(\omega)$ satisfy all the hypotheses listed in Problem 3.3.1 except (3.3.9). Further, assume that the elements $a_{ij}(t, \omega)$ are mutually independent. Define $L_{ij}(s, t) = E(\exp(sa_{ij}(t, \omega)))$. Then show that

(i)   if for some $s^* > 0$ and each $(i, j)$,

$$\int_{t_0}^t L_{ij}(s^*, t) \, dt \qquad \text{and} \qquad \int_{t_0}^t L_{ij}(-s^*, t) \, dt \tag{3.3.11}$$

are finite on $J$, then the sample solution of (3.3.8) is an $L^p$-solution, provided that

$$p < s^* n^2 a^{-1}; \tag{3.3.12}$$

(ii)   if for some $s^* > 0$ and each $(i, j)$,

$$\int_{t_0}^t (L_{ij}(s^*, t))^{n^2} dt \quad \text{and} \quad \int_{t_0}^t (L_{ij}(-s^*, t))^{n^2} dt \tag{3.3.13}$$

are finite on $J$, then the sample solution of (3.3.8) is an $L^p$-solution, provided that

$$p < s^* a^{-1}. \tag{3.3.14}$$

We note that if the moment-generating functions $L_{ij}(s, t)$ of $a_{ij}(t, \omega)$ in Problem 3.3.2 are defined everywhere as functions of $s$ and are locally $L^\infty$ functions of $t \in J$, then the sample solution of (3.3.8) is an $L^p$-solution for all $p$ for which $P(t)$ is $L^p$-integrable. This observation leads to the following problem.

**Problem 3.3.3.**   Suppose that $x_0$, $A(t, \omega)$, and $P(t, \omega)$ are as in Problem 3.3.2. Suppose further that the coefficients $a_{ij}(t, \omega)$ and $p_j(t, \omega)$ of $A(t, \omega)$ and $P(t, \omega)$, respectively, are normally distributed with means and variances that are bounded on $J$. Then show that the sample solution $x(t, \omega)$ of (3.3.8) is the $L^p$-solution for all $p$ on $J$.

**Problem 3.3.4.**   Consider a scalar random differential equation

$$x'(t, \omega) = a(\omega)x(t, \omega), \qquad x(t_0, \omega) = 1, \tag{3.3.15}$$

where $a \in R[\Omega, R]$ has the density function $\phi(\eta) = \frac{1}{2}\exp[-|\eta|]$. Then show that the sample solution of (3.3.15) $x(t, \omega)$ is an $L^p$-solution for $t \in [t_0, t_0 + (1/p))$.

   (*Hint*:   Note that

$$\int_{t_0}^t \left\| e^{(t-t_0)a(\omega)} \right\|_p ds = \int_{t_0}^t \left[ \int_{-\infty}^{\infty} e^{p(t-t_0)\eta} \phi(\eta) \, d\eta \right]^{1/p} ds$$

$$= \frac{(t - t_0)}{2} \int_{-\infty}^{\infty} \exp[p(t - t_0)\eta - |\eta|] \, d\eta < \infty,$$

provided that $t \in [t_0, t_0 + (1/p))$.)

The following Peano-type existence result gives the sufficient conditions under which every sample solution of (3.2.3) is also an $L^p$-solution.

**Theorem 3.3.2.** Assume that

($H_1$)  $f \in M[J \times \bar{B}(z, \rho), R[\Omega, R^n]]$ and $f(t, x, \omega)$ is sample continuous in $x$;

($H_2$)  $g \in M[J \times [0, 2\rho], R[\Omega, R_+]]$, $g(t, u, \omega)$ is sample continuous non-decreasing in $u$ for each $t \in J$ w.p. 1, and $g(t, u(t), \omega)$ is sample Lebesgue-integrable whenever $u(t)$ is sample absolutely continuous on $J$ and satisfies

$$\|f(t, x, \omega)\| \leq g(t, \|x\|, \omega) \tag{3.3.16}$$

a.e. on $J$ and for $x \in B(z, \rho)$;

($H_3$)  $u(t, \omega)$ is the sample maximal solution of

$$u'(t, \omega) = g(t, u(t, \omega), \omega), \qquad u(t_0, \omega) = u_0(\omega), \tag{3.3.17}$$

existing for $t \in J$, which is also an $L^p$-solution;

($H_4$)  $x_0(\omega) \in \bar{B}(z, \frac{1}{2}\rho)$ and $\|x_0(\omega)\| \leq \|u_0(\omega)\|$ w.p. 1.

Then the initial-value problem (3.2.3) has at least one $L^p$-solution $x(t, t_0, x_0)$ on $[t_0, t_0 + b]$ for some $b > 0$.

*Proof.* By Theorem 2.2.1 it follows that the initial-value problem (3.2.3) has at least one sample solution $x(t, t_0, x_0(\omega), \omega) = x(t, \omega)$ on $[t_0, t_0 + b]$ for some $b > 0$. By Lemma 2.2.1, we have

$$u(t, \omega) = u_0(\omega) + \int_{t_0}^t g(s, u(s, \omega), \omega)\, ds \qquad \text{on} \quad J$$

and

$$x(t, \omega) = x_0(\omega) + \int_{t_0}^t f(s, x(s, \omega), \omega)\, ds, \qquad t \in [t_0, t_0 + b],$$

where the integral denotes the sample integral. Using this, together with (3.3.17), ($H_2$), ($H_4$), and the scalar version of Theorem 2.5.1, we get

$$\|x(t, \omega)\| \leq u(t, \omega) \qquad \text{for} \quad t \in [t_0, t_0 + b]. \tag{3.3.18}$$

Now we show that $[x(t, \omega)] = x(t)$ is an $L^p$-solution of (3.2.3). By (3.3.18) it is obvious that $x(t) \in L^p[\Omega, R^n]$ on $[t_0, t_0 + b]$. From (3.3.16) and (3.3.17), we obtain

$$\|x(t, \omega) - x(t_1, \omega)\| \leq |u(t, \omega) - u(t_1, \omega)| \quad \text{w.p. 1} \tag{3.3.19}$$

for all $t_1, t \in [t_0, t_0 + b]$, which implies that

$$\|x(t) - x(t_1)\|_p \leq \|u(t) - u(t_1)\|_p \qquad \text{for all} \quad t, t_1 \in [t_0, t_0 + b].$$

This establishes the $p$th mean continuity of $x(t)$ because of the fact that $u(t)$ is $p$th mean continuous. Now we show that $x(t)$ has a $p$th mean derivative and satisfies $x'(t) = f(t, x(t))$, where $f(t, x(t)) = [f(t, x(t, \omega), \omega)]$.

For any $t \in [t_0, t_0 + b]$ and for any $h \in R$ such that $t + h \in [t_0, t_0 + b]$, we see, because of (3.3.19), that

$$\left\| \frac{x(t + h, \omega) - x(t, \omega)}{h} \right\| \leq \left| \frac{u(t + h, \omega) - u(t, \omega)}{h} \right| \quad \text{w.p. 1.} \quad (3.3.20)$$

We observe that

$$\frac{x(t + h, \omega) - x(t, \omega)}{h} \to f(t, x(t, \omega), \omega) \quad \text{w.p. 1} \quad (3.3.21)$$

a.e. on $[t_0, t_0 + b]$ and

$$\frac{u(t + h, \omega) - u(t, \omega)}{h} \to g(t, u(t, \omega), \omega) \quad \text{w.p. 1} \quad (3.3.22)$$

and in the sense of the $L^p$-integral a.e. on $[t_0, t_0 + b]$. From (3.3.20), (3.3.21), (3.3.22), and by the application of the generalized Lebesgue dominated convergence theorem (Theorem 3.1.2), we conclude that

$$\frac{x(t + h, \omega) - z(t, \omega)}{h} \to f(t, x(t, \omega), \omega)$$

in the sense of the $L^p$-integral a.e. on $[t_0, t_0 + b]$. This proves that $x(t) = [x(t, \omega)]$ has a $p$th mean derivative and satisfies the equation

$$x'(t) = f(t, x(t)) \quad \text{a.e. on} \quad [t_0, t_0 + b],$$

which proves the theorem.

**Problem 3.3.5.**   Let $g(t, u, \omega)$ in (3.3.17) be defined by

$$g(t, u, \omega) = K(t, \omega)u + p(t, \omega), \quad (3.3.23)$$

where $K, p \in M[J, R[\Omega, R_+]]$, $K(t, \omega)$ is sample Lebesgue-integrable on $J$ and $p(t, \omega)$ is $L^p$-integrable on $J$. Let $x_0 \in L^p[\Omega, R^n]$. Let $K(t, \omega)$, $p(t, \omega)$, and $x_0(\omega)$ be independent on $J$. Let $L(s, t) = E[\exp(sK(t, \omega))]$. Then show that if $L(s, t)$ and $p$ satisfy (3.3.9) and (3.3.10), respectively, (3.2.3) has an $L^p$-solution.

(*Hint*:   Apply Problem 3.3.1 and Theorem 3.3.2.)

**Remark 3.3.1.**   If $\rho = \infty$ in Theorems 3.3.1 and 3.3.2, then the corresponding results establish the global existence of solutions of $L^p$-problems. Also, based on the results on continuation of sample solutions of (3.2.3) in

Section 2.2, one can formulate the corresponding results covering the continuation of $L^p$-solutions of (3.2.3). We do not attempt this.

The following generalization of Theorem 3.3.1 is interesting in itself.

**Theorem 3.3.3.** Assume that all the hypotheses of Theorem 3.3.2 are satisfied. Further, assume that $f$ satisfies the relation

$$\|f(t, x, \omega) - f(t, y, \omega)\| \le G(t, \|x - y\|, \omega) \quad \text{w.p. 1} \qquad (3.3.24)$$

a.e. on $J$ and $x, y \in \bar{B}(z, \rho)$, where $G \in M[J \times [0, 2\rho], R[\Omega, R_+]]$, $G(t, u, \omega)$ is sample continuous is $u$ for almost all $t \in J$ w.p. 1, and $G(t, u(t), \omega)$ is sample Lebesgue-integrable on $J$ whenever $u(t)$ is sample absolutely continuous on $J$. Let $u(t) \equiv 0$ be the unique $L^p$-solution of

$$u'(t, \omega) = G(t, u(t, \omega), \omega), \qquad u(t_0, \omega) = 0, \qquad (3.3.25)$$

existing on $J$. Then the differential system (3.2.3) has a unique $L^p$-solution on $J$.

*Proof.* By Theorem 3.3.2, the initial-value problem (3.2.3) has an $L^p$-solution. Let $x(t, \omega) = x(t, t_0, x_0, \omega)$ and $y(t, \omega) = y(t, t_0, x_0, \omega)$ be two $L^p$-solutions of (3.2.3) on $J$. Set

$$m(t, \omega) = \|x(t, \omega) - y(t, \omega)\| \qquad \text{so that} \quad m(t_0, \omega) = 0.$$

From (3.3.25) we have

$$m'(t, \omega) \le \|f(t, x(t, \omega), \omega) - f(t, y(t, \omega), \omega)\|$$
$$\le G(t, m(t, \omega), \omega) \quad \text{w.p. 1.}$$

By Theorem 2.5.1 we then obtain

$$m(t, \omega) \le r(t, \omega), \qquad t \in [t_0, t_0 + a], \qquad (3.3.26)$$

where $r(t, \omega)$ is the maximal solution of (3.3.25) through $(t_0, 0)$. From (3.3.26) we get

$$\|m(t)\|_p \le \|r(t)\|_p \qquad \text{on} \quad J.$$

By assumption, $r(t) \equiv 0$ on $J$, and so $m(t) \equiv 0$ on $J$. This completes the proof.

**Remark 3.3.2.** The function $G(t, u, \omega) = K(t, \omega)u$ is admissible in Theorem 3.3.3, where $K \in M[J, R[\Omega, R_+]]$ and $K(t, \omega)$ is sample Lebesgue-integrable on $J$. In this case, Theorem 3.3.3 reduces to Theorem 3.3.1.

### 3.4. CONTINUOUS DEPENDENCE

This section deals with the continuous dependence of solutions on the parameter.

**Theorem 3.4.1.**   Assume that

$(H_1)_\lambda$   $f \in M[J \times B(z, \rho) \times \Lambda, R[\Omega, R^n]]$ and $f(t, x, \lambda, \omega)$ is sample continuous in $x$ for each $(t, \lambda)$ and continuous in $(x, \lambda)$ for each fixed $t$, $\Lambda$ being an open $\lambda$-set in $R^m$;

$(H_2)_\lambda$   $G \in M[J \times R_+, R[\Omega, R_+]]$, $G(t, u, \omega)$ is sample continuous nondecreasing in $u$ for each $t \in J$, and $G(t, u(t), \omega)$ is sample Lebesgue-integrable whenever $u(t)$ is sample absolutely continuous;

$(H_3)_\lambda$   $u(t) \equiv 0$ is the unique $L^p$-solution of (3.3.25) such that $u(t_0) = 0$ and the solutions $u(t, t_0, u_0, \omega)$ of (3.3.25) are $L^p$-continuous with respect to $(t_0, u_0)$;

$(H_4)_\lambda$   for $(t, x, \lambda), (t, x, \lambda) \in J \times B(z, \rho) \times \Lambda$,

$$\left\| f(t, x, \lambda, \omega) - f(t, y, \lambda, \omega) \right\| \leq G(t, \|x - y\|, \omega), \text{ w.p. 1};   (3.4.1)$$

$(H_5)$   there exist product-measurable functions $K, p \in M[J, R[\Omega, R_+]]$ such that $K(t, \omega)$ is sample Lebesgue-integrable on $J$ and $p(t, \omega)$ is $L^p$-integrable on $J$; also, let $x_0(\lambda, \omega) \in L^p[\Omega, R^n]$ be $L^p$-continuous in $\lambda$ and $K(t, \omega)$, $p(t, \omega)$, $x_0(\lambda, \omega)$ be independent; furthermore, suppose that

$$\left\| f(t, x, \omega) \right\| \leq K(t, \omega)\|x\| + p(t, \omega)   \text{w.p. 1}   (3.4.2)$$

and

$$\int_{t_0}^t \left\| K(s, \omega) \int_{t_0}^{t+a} \left\| f(v, x_0(\lambda, \omega), \lambda, \omega) \right\| dv \exp\left[ \int_{t_0}^s K(u, \omega)\, dn \right] \right\|_p ds < \infty   (3.4.3)$$

on $J$;

$(H_6)_\lambda$   $f(t, x, \lambda, \omega)$ is $L^p$-continuous in $\lambda \in \Lambda$.

Then given $\varepsilon > 0$, there exists a $\delta(\varepsilon) > 0$ such that for every $\lambda, \|\lambda - \lambda_0\| < \delta(\varepsilon)$, and the system of differential equations

$$x'(t, \omega) = f(t, x(t, \omega), \lambda, \omega),   x(t_0, \lambda, \omega) = x_0(\lambda, \omega)   (3.4.4)$$

admits unique $L^p$-solution $x(t, \omega) = x(t, t_0, x_0, \lambda, \omega)$ satisfying

$$\left\| x(t, \lambda, \omega) - x(t, \lambda_0, \omega) \right\|_p < \varepsilon   \text{for}   t \in J.   (3.4.5)$$

*Proof.*   Under hypotheses $(H_1)$ and $(H_5)$, Theorem 3.3.2 shows that the initial-value problem (3.4.4) has at least one $L^p$-solution $x(t, \lambda, \omega) = x(t, t_0, x_0(\lambda, \omega), \omega)$ on $[t_0, t_0 + b]$. Furthermore, because of $(H_2)_\lambda$–$(H_4)_\lambda$ and Theorem 3.3.3, $x(t, \lambda)$ is the unique $L^p$-solution process of (3.4.4).

To prove that (3.4.5) holds, let $x(t, \lambda)$ and $x(t, \lambda_0)$ be $L^p$-solutions of (3.4.4) with $\|\lambda - \lambda_0\| < \delta(\varepsilon)$. Set

$$m(t, \omega) = \left\| x(t, \lambda, \omega) - x(t, \lambda_0, \omega) \right\|$$

so that

$$m(t_0, \omega) = \left\| x_0(\lambda, \omega) - x_0(\lambda_0, \omega) \right\|.$$

It is obvious that

$$m(t, \omega) \leq \left\| x_0(\lambda, \omega) - x_0(\lambda_0, \omega) \right\|$$

$$+ \int_{t_0}^{t} \left[ \left\| f(s, x(s, \lambda, \omega), \lambda, \omega) - f(s, x(s, \lambda_0, \omega), \lambda_0, \omega) \right\| \right] ds$$

$$\leq m(t_0, \omega) + \int_{t_0}^{t} \left[ \left\| f(s, x(s, \lambda, \omega), \lambda, \omega) - f(s, x(s, \lambda_0, \omega), \lambda, \omega) \right\| \right] ds$$

$$+ \int_{t_0}^{t} \left\| f(s, x(s, \lambda_0, \omega), \lambda, \omega) - f(s, x(s, \lambda_0, \omega), \lambda_0, \omega) \right\| ds$$

$$\leq \eta + \int_{t_0}^{t} G(s, m(s, \omega), \omega) \, ds, \qquad\qquad (3.4.6)$$

where

$$\eta(\omega) = m(t_0, \omega) + \int_{t_0}^{t_0+b} \left\| (s, x(s, \lambda_0, \omega), \lambda, \omega) - f(s, x(s, \lambda_0, \omega), \lambda_0, \omega) \right\| ds.$$

Setting

$$u(t, \omega) = \eta + \int_{t_0}^{t} G(s, m(s, \omega), \omega), \qquad u(t_0, \omega) = \eta(\omega),$$

relation (3.4.6), together with the nondecreasing property of $G$, yields

$$u'(t, \omega) \leq G(t, u(t, \omega), \omega), \qquad u(t_0, \omega) = \eta(\omega).$$

Hence by Theorem 2.5.1 we have

$$u(t, \omega) \leq r(t, t_0, u_0, \omega), \qquad t \in [t_0, t_0 + b],$$

where $r(t, t_0, u_0, \omega)$ is the maximal solution of (3.3.25) through $(t_0, u_0)$ with $u_0 = \eta$. This, together with (3.4.6), gives

$$m(t) \leq r(t, t_0, \eta), \qquad t \in [t_0, t_0 + b].$$

Since the solutions $u(t, t_0, u_0, \omega)$ of (3.3.25) are assumed to be $L^p$-continuous with respect to $u_0$, it follows that

$$\lim_{\eta \to 0} r(t, t_0, \eta) = r(t, t_0, 0)$$

and, by hypotheses, $r(t, t_0, 0) \equiv 0$. This, in view of the definition of $m(t, \omega)$, $(H_6)_\lambda$, and $\eta$, yields

$$\lim_{\lambda \to \lambda_0} \left\| x(t, \lambda, \omega) - x(t, \lambda_0, \omega) \right\|_p = 0,$$

which establishes the theorem.

We are now ready to prove a theorem that establishes the continuous dependence of solutions of (3.2.3) on the initial conditions $(t_0, x_0)$.

**Theorem 3.4.2.** Assume that the hypotheses of Theorem 3.3.3 hold. Furthermore, assume that the solution process $u(t, t_0, u_0, \omega)$ is $L^p$-continuous with respect to its initial data. Then the solutions $x(t, t_0, x_0, \omega)$ of (3.2.3) through $(t_0, x_0)$ are $L^p$-unique and $L^p$-continuous with respect to the initial conditions $(t_0, x_0)$.

*Proof.* The proof easily follows from the proofs of Theorems 2.6.3 and 3.4.1. We omit the details.

We remark that the $L^p$-differentiability of solutions with respect to initial conditions can be studied analogously to the differentiability of solutions of abstract differential systems. See Ladas and Lakshmikantham [49]. Moreover, based on the previous results, results similar to those of Section 2.7 can be developed. We omit such results to avoid monotony.

### 3.5.  COMPARISON THEOREMS

Consider the system of first-order differential equations

$$x' = f(t, x), \qquad x(t_0) = x_0, \tag{3.5.1}$$

where the prime denotes the $p$th mean derivative of $x$, $f \in C[R_+ \times B(x_0, \rho)$, $L^p[\Omega, R^n]]$, and $f(t, x)$ is smooth enough to guarantee the existence of $L^p$-solutions $x(t) = x(t, t_0, x_0)$ of (3.5.1) for $t \geq t_0$, $t_0 \in R_+$.

The following results give estimates for the solutions of (3.5.1) in terms of maximal solutions of a deterministic comparison system.

**Theorem 3.5.1.** Assume that

(i)   $g \in C[R_+ \times R^m, R^m]$ and $g(t, u)$ is quasi-monotone nondecreasing in $u$ for fixed $t \in R_+$;

(ii)   $r(t) = r(t, t_0, u_0)$ is the maximal solution of a deterministic differential system

$$u' = g(t, u), \qquad u(t_0) = u_0, \tag{3.5.2}$$

existing for $t \geq t_0$;

(iii)   $V \in C[R_+ \times L^p[\Omega, R^n], R^m]$, $V(t, x)$ satisfies a local $L^p$-Lipschitz condition in $x$, and for $(t, x) \in R_+ \times L^p[\Omega, R^n]$,

$$D^+ V(t, x) \equiv \limsup_{h \to 0^+} \frac{1}{h}[V(t + h, x + hf(t, x)) - V(t, x)]$$

$$\leq g(t, V(t, x)); \tag{3.5.3}$$

(iv)   $x(t)$ is any $L^p$-solution of (3.5.1), existing for $t \geq t_0$, such that

$$V(t_0, x_0) \leq u_0. \tag{3.5.4}$$

Then

$$V(t, x(t)) \leq r(t) \qquad \text{for} \quad t \geq t_0. \tag{3.5.5}$$

*Proof.*   Set

$$m(t) = V(t, x(t)) \qquad \text{so that} \quad m(t_0) = V(t_0, x_0),$$

where $x(t)$ is any $L^p$-solution of (3.5.1). For small $h > 0$, we have

$$
\begin{aligned}
m(t + h) - m(t) &= V(t + h, x(t + h)) - V(t, x(t)) \\
&= V(t + h, x(t) + hf(t, x(t))) - V(t, x(t)) \\
&\quad + V(t + h, x(t + h)) - V(t + h, x(t) + hf(t, x(t))).
\end{aligned}
$$

Hypothesis (iii) now yields the inequality

$$D^+ m(t) \leq g(t, m(t)), \qquad t \geq t_0. \tag{3.5.6}$$

By Theorem 1.7.1, we then have

$$m(t) \leq r(t) \qquad \text{for} \quad t \geq t_0,$$

and the proof is complete.

**Corollary 3.5.1.**   Assume that the hypotheses of Theorem 3.5.1 hold with $g(t, u) \equiv 0$. Then the function $V(t, x(t))$ is nondecreasing in $t$, and

$$V(t, x(t)) \leq V(t_0, x_0) \qquad \text{for} \quad t \geq t_0.$$

**Theorem 3.5.2.**   Let the hypotheses of Theorem 3.5.1 hold except that inequality (3.5.3) is strengthened to

$$A(t)D^+ V(t, x) + A'(t)V(t, x) \leq g(t, A(t)V(t, x)) \tag{3.5.7}$$

for $(t, x) \in R_+ \times L^p[\Omega, R^n]$, where $A(t)$ is a continuously differentiable $m \times m$ matrix function such that $A^{-1}(t)$ exists and its elements are nonnegative and the elements of the matrix $A^{-1}(t)A(t) = (\alpha_{ij}(t))$ satisfy $\alpha_{ij}(t) \leq 0$ for $i \neq j$. Then

$$A(t_0)V(t_0, x_0) \leq u_0 \tag{3.5.8}$$

implies

$$V(t, x(t)) \leq R(t), \qquad t \geq t_0, \tag{3.5.9}$$

where $R(t) = R(t, t_0, v_0)$ is the maximal solution of the differential system

$$v' = A^{-1}(t)[-A'(t)v + g(t, A(t)v)], \qquad v(t_0) = v_0, \tag{3.5.10}$$

existing for $t \geq t_0$, and $A(t_0)v_0 = u_0$.

*Proof.*   Set $W(t, x) = A(t)V(t, x)$. From (3.5.7) and the nature of $A(t)$, it is obvious that

$$D^+ W(t, x) \leq g(t, W(t, x))$$

for $(t, x) \in R_+ \times L^p[\Omega, R^n]$. Now by following the argument used in the proof of Theorem 2.8.2, it is easy to conclude the proof of the theorem.

## 3.6.   STABILITY CRITERIA

Let $x(t) = x(t, t_0, x_0)$ be any solution of (3.5.1). Without loss of generality, we assume that $x(t) \equiv 0$ is the unique solution of (3.5.1) through $(t_0, 0)$.

We list a few definitions concerning the stability of the trivial solution of (3.5.1).

**Definition 3.6.1.**   The trivial solution of (3.5.1) is said to be

($S_1$)   *stable* if for each $\varepsilon > 0$ and $t_0 \in R_+$, there exists a positive function $\delta = \delta(t_0, \varepsilon)$ that is continuous in $t_0$ for each $\varepsilon$ such that $\|x_0\|_p < \delta$ implies $\|x(t)\|_p < \varepsilon$ for $t \geq t_0$;

($S_2$)   *uniformly stable* if ($S_1$) holds with $\delta$ independent of $t_0$;

($S_3$)   *asymptotically stable* if it is stable and if for any $\varepsilon > 0$ and $t_0 \in R_+$, there exist positive numbers $\delta_0 = \delta_0(t_0)$ and $T = T(t_0, \varepsilon)$ such that $\|x_0\|_p \leq \delta$ implies $\|x(t)\|_p < \varepsilon$ for $t \geq t_0 + T$;

($S_4$)   *uniformly asymptotically stable* if ($S_1$) and ($S_3$) hold with $\delta$, $\delta_0$, and $T$ independent of $t_0$.

It is obvious that the stability, uniform stability, asymptotic stability, and uniform asymptotic stability of the trivial solution of (3.5.1) are equivalent to ($SM_1$)–($SM_4$), respectively, in Definition 2.10.1.

Let us assume that the comparison equation (3.5.2) possesses the trivial solution. Then we can define the corresponding stability concepts ($S_1^*$)–($S_4^*$) for the trivial solution of (3.5.2). For example, ($S_1^*$) implies the following definition:

**Definition 3.6.2.**   The trivial solution $u \equiv 0$ of (3.5.2) is said to be ($S_1^*$) *stable* if for each $\varepsilon > 0$ and $t_0 \in R_+$, there exists a positive function $\delta = \delta(t_0, \varepsilon)$ that is continuous in $t_0$ for each $\varepsilon$ such that $\sum_{i=1}^m u_{i0} \leq \delta$ implies $\sum_{i=1}^m u_i(t, t_0, u_0) < \varepsilon$ for $t \geq t_0$. The definitions of ($S_2^*$)–($S_4^*$) can be formulated similarly.

We now present a result concerning the stability of the trivial solution of (3.5.1)

**Theorem 3.6.1.**   Assume that

(i)   $g \in C[R_+ \times R^m, R^m]$ and $g(t, u)$ is quasi-monotone nondecreasing in $u$ for fixed $t \in R_+$;

(ii)   $V \in C[R_+ \times B(\rho), R^m]$, $V(t, x)$ satisfies a local $L^p$-Lipschitz condition in $x$, and for $(t, x) \in R_+ \times S(\rho)$,

$$D^+_{(3.5.1)}V(t, x) \le g(t, V(t, x)), \qquad (3.6.1)$$

where $B(\rho) = \{x \in L^p[\Omega, R^n] : \|x\|_p < \rho\}$;
(iii)   for $(t, x) \in R_+ \times S(\rho)$

$$b(\|x\|_p) \le \sum_{i=1}^{m} V_i(t, x) \le a(t, \|x\|_p), \qquad (3.6.2)$$

where $b, a(t, \cdot) \in \mathcal{K}$ and $a \in C[R_+ \times R_+, R_+]$. Then

$$(S_1^*) \quad \text{implies} \quad (S_1).$$

*Proof.*   Let $\varepsilon > 0$ and $t_0 \in R_+$ be given. Assume that $(S_1^*)$ holds. Then given $b(\varepsilon) > 0$ and $t_0 \in R_+$, there exists a positive function $\delta_1 = \delta_1(t_0, \varepsilon)$, continuous in $t_0$ for each $\varepsilon$, such that $\sum_{i=1}^{m} u_{i0} \le \delta$ implies

$$\sum_{i=1}^{m} u_i(t, t_0, u_0) < b(\varepsilon), \qquad t \ge t_0. \qquad (3.6.3)$$

Let us choose $u_0 = (u_{10}, u_{20}, \ldots, u_{n0})^T$ so that $V(t_0, x_0) \le u_0$ and $\sum_{i=1}^{m} u_{i0} = a(t_0, \|x_0\|_p)$ for $x_0 \in B(\rho)$. Since $a(t_0, \cdot) \in K$ and $a \in C[R_+ \times R_+, R_+]$, we can find a $\delta = \delta(t_0, \varepsilon)$ that is continuous in $t_0$ for each $\varepsilon$ such that $\|x_0\|_p \le \delta$ implies $a(t_0, \|x_0\|_p) < \delta_1$. We claim that this $\delta$ is good for $(S_1)$, that is, $\|x_0\|_p \le \delta$ implies that $\|x(t, t_0, x_0)\|_p < \varepsilon$, $t \ge t_0$. In fact, from relations (3.6.2), (3.5.5), and (3.6.3), we get

$$b(\|x(t, t_0, x_0)\|_p) \le V(t, x(t, t_0, x_0)) \le r(t, t_0, u_0) < b(\varepsilon)$$

for $t \ge t_0$. Since $b(r)$ is strictly increasing in $r$

$$\|x(t, t_0, x_0)\|_p < \varepsilon, \qquad t \ge t_0.$$

Therefore, $(S_1)$ holds.

**Problem 3.6.1.**   Assuming the hypotheses of Theorem 3.6.1, prove that $(S_3^*)$ implies $(S_3)$.

**Problem 3.6.2.**   Let the hypotheses of Theorem 3.6.1 be satisfied with $a(t, r) \equiv a(r)$. Then show that

(a)   $(S_2^*)$ implies $(S_2)$;
(b)   $(S_4^*)$ implies $(S_4)$.

We remark that one can formulate the stability results with regard to the trivial solution of (3.5.1) and its perturbed system

$$y' = f(t, y) + F(t, y), \qquad y(t_0) = x_0, \qquad (3.6.4)$$

by means of the method of variation of parameters. The formulation and the proofs of these results can be given on the basis of the results in Sections 2.11–2.13 and the results in Ladas and Lakshmikantham [49]. Such a detailed formulation and the proofs are left to the reader as exercises.

### Notes

The contents of Section 3.1 are based on well-known books, namely, Dunford and Schwartz [15], Hille and Phillips [25], and also Ladas and Lakshmikantham [49]. The results of Section 3.2 are taken from Strand [83, 84]. See also Edsinger [16]. Theorem 3.3.1 is adapted from Strand [84]. All the examples and problems given in Section 3.3 are taken from Strand [83]. See also Bharucha-Reid [5] and Tsokos and Padgett [85, 86]. Theorems 3.3.2 and 3.3.3 are new. All the results of Section 3.4 are new and are based on the results in Ladas and Lakshmikantham [49]. The results in Section 3.5 are formulated in the framework of Lakshmikantham and Leela [59]. In Section 3.6 sufficient conditions are given for the stability of the trivial solution of the $L^p$-initial-value problem in the context of the vector Lyapunov functions and the theory of deterministic differential inequalities.

# CHAPTER 4

# ITÔ–DOOB CALCULUS APPROACH

## 4.0. INTRODUCTION

Several phenomena in biological, physical, and social sciences can also be described by a system of stochastic differential equations of Itô-type,

$$dx = f(t, x)\,dt + \sigma(t, x)\,dz, \qquad x(t_0) = x_0, \tag{4.0.1}$$

where $dx$ is a stochastic differential of $x$, $z$ an $m$-dimensional normalized Wiener process defined on a complete probability space $(\Omega, \mathscr{F}, P)$, $f(t, x)$ an average rate vector, and $\sigma(t, x)$ a diffusion rate matrix function. This chapter concentrates on the study of Itô-type systems of the form (4.0.1). In Section 4.1 we sketch Itô's calculus. We begin with basic definitions and arrive at the definition of Itô's integral. We state important properties of stochastic integrals, introduce the stochastic differential, and derive a well-known differential formula known as Itô's formula. Some useful applications of Itô's formula are also given. Finally, a special class of Markov processes is discussed, and the Itô-type stochastic differential equation is formulated. Certain properties of diffusion processes are also stated. A Peano-type existence theorem for Itô-type stochastic differential systems is proved in Section 4.2. Several existence and uniqueness theorems are also developed. Continuous dependence on parameters, as well as continuity and differentiability properties of solutions with respect to initial data, is discussed in Section 4.3. Section 4.4 develops variation of constants formulas with respect to both stochastic and deterministic perturbations. Some basic results on stochastic differential inequalities of Itô-type are given in Section 4.5. The notion of maximal and minimal solutions relative to (4.0.1) is introduced in Section 4.6. In Section 4.7 basic comparison theorems for (4.0.1) are listed. By using the concept of Lyapunov function and the theory of stochastic and deterministic differential inequalities, several comparison theorems are developed in Section 4.8. Finally, in Sections 4.9, 4.10, and 4.11 a variety of stability results concerning stability in probability, stability in the $p$th mean, and stability with probability one of the trivial solution of (4.0.1) are presented in a coherent way. These results are developed in the framework of vector Lyapunov functions and the theory of differential inequalities as well as variation of constants formulas. This approach provides stabilizing as well as destabilizing effects of random disturbances. Moreover, the stability conditions are formulated in the context of the laws of large numbers.

### 4.1.  ITÔ'S CALCULUS

In this section we shall discuss Itô-type stochastic integral and differential calculus.

Let $\Omega \equiv (\Omega, \mathscr{F}, P)$ be a complete probability space. Let $\mathscr{F}_t$ be a sub-$\sigma$-algebra of $\mathscr{F}$ defined on $R_+$. Let $z(t)$ be an $m$-dimensional normalized Wiener process, and let $\mathscr{Z}_t$ and $\mathscr{Z}_t^+$ be the smallest sub-$\sigma$-algebras of $\mathscr{F}$ generated by all random variables $z(t)$ and $z(s) - z(t)$ for $s \geq t \geq 0$, respectively. We need the following definitions.

**Definition 4.1.1.**  A family $\{\mathscr{F}_t : t \in R_+\}$ of sub-$\sigma$-algebras of $\mathscr{F}$ is said to be nonanticipating with respect to the $m$-dimensional normalized Wiener process if it satisfies the following conditions:

(i)    $\mathscr{F}_s \subset \mathscr{F}_t$ for all $t > s \geq 0$;
(ii)   $\mathscr{Z}_t \subset \mathscr{F}_t$ for all $t \in R_+$;
(iii)  $\mathscr{F}_t$ is independent of $\mathscr{Z}_t^+$.

Let us note that the family $\{\mathscr{Z}_t : t \geq 0\}$ is the smallest nonanticipating family of sub-$\sigma$-algebras of $\mathscr{F}$.

**Definition 4.1.2.**  An $n \times m$ product-measurable matrix function $G(t, \omega)$ defined on $R^+ \times \Omega$ into $R^{nm}$ is said to be nonanticipating with respect to a nonanticipating family $\{\mathscr{F}_t : t \in R_+\}$ of sub-$\sigma$-algebras of $\mathscr{F}$ if $G(t, \omega)$ is $\mathscr{F}_t$-measurable.

Let $M_2[a, b]$ denote the set of all nonanticipating $n \times m$ matrix functions $G = G(t, \omega)$ defined on $[a, b] \times \Omega$ into $R^{nm}$ such that

$$\int_a^b \|G(s, \omega)\|^2 \, ds < \infty \quad \text{w.p. 1,}$$

where the integral is the sample Lebesgue integral.

We note that every deterministic function $G(t, \omega) \equiv G(t)$ is a nonanticipating function. Also, if $G$ is nonanticipating, then every product-measurable function $g(t, G)$ defined on $R^+ \times R^{nm}$ into $R^n$ is nonanticipating.

We are now in a position to define the stochastic integral of a step function $G(t, \omega) \in M_2[a, b]$ relative to $z(t)$.

**Definition 4.1.3.** Let $G \in M_2[a, b]$ be a step function, and let $a = t_0 < t_1 < \cdots < t_n = b$ be a partition of the interval $[a, b]$ such that $G(t, \omega) = G(t_i, \omega)$ for $t \in [t_i, t_{i+1})$, $i = 1, 2, \ldots, n - 1$. The stochastic integral of $G$ with respect to an $m$-dimensional Wiener process $z(t)$ is then defined by

$$\int_a^b G(t) \, dz(t) = \sum_{i=0}^{n-1} G(t_i)[z(t_{i+1}) - z(t_i)], \tag{4.1.1}$$

where

$$\int_a^b G(t) \, dz(t) = \int_a^b G(t, \omega) \, dz(t, \omega)$$

and

$$\sum_{i=0}^{n-1} G(t_i)[z(t_{i+1}) - z(t_i)] = \sum_{i=0}^{n-1} G(t_i, \omega)[z(t_{i+1}, \omega) - z(t_i, \omega)].$$

We list below some needed properties of this stochastic integral:

(a)   If $G_1, G_2 \in M_2[a, b]$ are step functions, then

$$\int_a^b (\alpha G_1(t) + \beta G_2(t)) \, dz(t) = \alpha \int_a^b G_1(t) \, dz(t) + \beta \int_a^b G_2(t) \, dz(t), \tag{4.1.2}$$

where $\alpha, \beta \in R$. If $G \in M_2[a, b]$ is a step function, then

(b)   $\int_a^b G(t) \, dz(t) = \left( \sum_{j=1}^{m} \int_a^b G_{1j}(t) \, dz_j(t), \ldots, \sum_{j=1}^{m} \int_a^b G_{nj}(t) \, dz_j(t) \right)^{\mathrm{T}}$;   (4.1.3)

(c)   $E[\|G(t)\|] < \infty$, $t \in [a, b]$, implies

$$E\left[ \int_a^b G(t) \, dz(t) \right] = 0;$$

(d)   $E[\|G(t)\|^2] < \infty$, $t \in [a, b]$, implies,

$$E\left[ \left( \int_a^b G(t) \, dz(t) \right) \left( \int_a^b G(t) \, dz(t) \right)^{\mathrm{T}} \right] = \int_a^b E[G(t)G^{\mathrm{T}}(t)] \, dt;$$

and, in particular,

$$E\left[ \left\| \int_a^b G(t) \, dz(t) \right\|^2 \right] = \int_a^b E[\|G(t)\|^2] \, dt;$$

(e)   for all $\varepsilon > 0$ and $c > 0$,

$$P\left\{ \left\| \int_a^b G(t) \, dz(t) \right\| > c \right\} \leq \frac{\varepsilon}{c^2} + P\left\{ \int_a^b \|G(t)\|^2 \, dt > \varepsilon \right\}.$$

The following result provides a motivation for the stochastic integral of an arbitrary $n \times m$ matrix function $G(t, \omega)$ that belongs to $M_2[a, b]$.

**Lemma 4.1.1.**   Let $G \in M_2[a, b]$, and let $G_n(t, \omega)$ be a sequence of $n \times m$ matrix step functions such that $G_n \in M_2[a, b]$ and

$$\int_a^b \|G(t, \omega) - G_n(t, \omega)\|^2 \, dt \to 0 \qquad (4.1.4)$$

in probability as $n \to \infty$. Then the random sequence $\int_a^b G_n(t) \, dz(t)$ converges in probability to some limit. Moreover, this limit is independent of the selection of a sequence.

*Proof.*   From (4.1.4) it follows that

$$\int_a^b \|G_n(t) - G_m(t)\|^2 \, dt \to 0$$

in probability as $n, m \to \infty$. Applying property (e) to the step function $G_n(t) - G_m(t)$, we get

$$\limsup_{n,m \to \infty} P\left\{ \left\| \int_a^b G_n(t) \, dz(t) - \int_a^b G_m(t) \, dz(t) \right\| > c \right\}$$

$$\leq \frac{\varepsilon}{c^2} + \limsup_{n,m \to \infty} P\left\{ \int_a^b \|G_n(t) - G_m(t)\|^2 \, dt > \varepsilon \right\} = \frac{\varepsilon}{c^2}.$$

Since $\varepsilon$ is arbitrary, we have

$$\lim_{n,m \to \infty} P\left\{ \left\| \int_a^b G_n(t) \, dz(t) - \int_a^b G_m(t) \, dz(t) \right\| > c \right\} = 0.$$

This implies that $\int_a^b G_n(t) \, dz(t)$ is a Cauchy sequence in probability and hence converges in probability to an $n$-random vector $I(G)$. The limit is almost-surely uniquely determined and independent of the special choice of sequences for which (4.1.4) holds.

We are now ready to define the stochastic integral.

**Definition 4.1.4.**   Let $G \in M_2[a, b]$, and let $G_n(t, \omega)$ be a sequence of $n \times m$ matrix step functions such that $G_n \in M_2[a, b]$ and

$$\int_a^b \|G(t, \omega) - G_n(t, \omega)\|^2 \, dt \to 0$$

in probability as $n \to \infty$. The stochastic integral or Itô's integral of $G$ with respect to an $m$-dimensional Wiener process $z(t)$ is defined by

$$\int_a^b G(t) \, dz(t) = \lim_{n \to \infty} \int_a^b G_n(t) \, dz(t) \quad \text{in probability.}$$

The stochastic integral of $G \in M_2[a, b]$ possesses the following properties:

(I)   It possesses properties (a), (b), and (e) of the stochastic integral of an $n \times m$ matrix step function that belongs to $M_2[a, b]$.

(II)   If $\int_a^b E[\|G(t)\|^2]\,dt < \infty$, then
   (i)   there exists a sequence $G_n \in M_2[a, b]$ of step functions such that

$$\int_a^b E[\|G_n(t)\|^2]\,dt < \infty,$$

$$\lim_{n \to \infty} \int_a^b E[\|G_n(t) - G(t)\|^2]\,ds = 0$$

and

$$\lim_{n \to \infty} \int_a^b G_n(t)\,dz(t) = \int_a^b G(t)\,dz(t)$$

in the mean square sense;

(ii)   $E\left[\int_a^b G(t)\,dz(t)\right] = 0;$  (4.1.5)

(iii)   $E\left[\left(\int_a^b G(t)\,dz(t)\right)\left(\int_a^b G(t)\,dz(t)\right)^{\mathrm{T}}\right] = \int_a^b E[G(t)G^{\mathrm{T}}(t)]\,dt;$  (4.1.6)

(iv)   $E\left[\left\|\int_a^b G(t)\,dz(t)\right\|^2\right] = \int_a^b E[\|G(t)\|^2]\,dt.$  (4.1.7)

The following example is instructive.

**Example 4.1.1.**   Let $z(t)$ be a scalar normalized Wiener process. Since $z(t)$ is sample continuous, it follows that $z \in M_2[a, b]$. Then

$$\sum_{i=0}^{n-1} z(t_i)[z(t_{i+1}) - z(t_i)] \xrightarrow{i.p.} \int_a^b z(t)\,dz(t)$$

as $\delta_n \to 0$, where $\delta_n = \max(t_{i+1} - t_i)$, $t_0 = a < t_1 < \cdots < t_n = b$. Moreover,

$$\sum_{i=0}^{n-1} z(t_i)[z(t_{i+1}) - z(t_i)] \xrightarrow{m.s.} \tfrac{1}{2}(z^2(b) - z^2(a)) - \tfrac{1}{2}(b - a),$$

and

$$\int_a^b z(t)\,dz(t) = \tfrac{1}{2}[z^2(b) - z^2(a) - (b - a)],$$  (4.1.8)

where m.s. stands for mean square.

The result contained in the next example is important in many situations.

**Example 4.1.2.**   Suppose that an $n \times m$ a.s. sample continuous random matrix function $G$ belongs to $M_2[a, b]$. Then

$$\int_a^b G(t)\,dz(t) = \lim_{\delta_n \to 0} \sum_{i=0}^{n} G(t_i)[z(t_{i+1}) - z(t_i)]$$

in probability, where $\delta_n = \max\{(t_{i+1} - t_i)\}$ and $t_0 = a < t_1 < t_2 < \cdots < t_{n+1} = b$.

*Proof.* The sequence of step functions $G_n(t) = \sum_{i=0}^n G(t_i)I_{[t_i,t_{i+1}]}(t)$, where $I_{[t_i,t_{i+1}]}(t)$ is a characteristic function relative to $[t_i, t_{i+1}]$. Because of the a.s. sample continuity of $G$, we get

$$\lim_{n \to \infty} \int_a^b \|G(t) - G_n(t)\|^2 \, dt = 0 \quad \text{in probability.}$$

Therefore, by the definition of the stochastic integral, we have

$$\lim_{n \to \infty} \int_a^b G_n(t) \, dz(t) = \int_a^b G(t) \, dz(t)$$

in probability, which is equivalent to

$$\lim_{\delta_n \to 0} \sum_{i=0}^n G(t_i)[z(t_{i+1}) - z(t_i)] = \int_a^b G(t) \, dz(t)$$

in probability.

We shall next define the stochastic indefinite integral for a matrix function $G \in M_2[a, b]$. Let $[a, t] \subset [a, b]$, and let $I_{[a,t]}$ be a characteristic function of $[a, t]$. It is obvious that $GI_{[a,t]} \in M_2[a, b]$.

**Definition 4.1.5.** The stochastic indefinite integral of an $n \times m$ matrix function $G \in M_2[a, b]$ is defined by

$$x(t) \equiv \int_a^t G(s) \, dz(s) = \int_a^b G(s)I_{[a,t]}(s) \, dz(s), \tag{4.1.9}$$

which is an $n$-dimensional stochastic process defined (uniquely up to stochastic equivalence) for all $t \in [a, b]$.

The indefinite integral $x(t)$ of (4.1.9) possesses the following properties. Without loss of generality, we shall assume that $x(t)$ is a separable process.

(a)   $x(a) = 0$ w.p. 1.
(b)   $x(t) - x(s) = \int_s^t G(u) \, dz(u)$.
(c)   $x(t)$ is a random vector relative to $(\Omega, \mathscr{F}_t, P)$.
(d)   If $\int_a^t E[\|G(s)\|^2] \, ds < \infty$ for all $t \in [a, b]$, then $\{x(t), \mathscr{F}_t : t \in [a, b]\}$ is a martingale (see Appendix). Furthermore, for $t, s \in [a, b]$, we have

$$E[x(t)] = 0, \tag{4.1.10}$$

$$E[x(t)x^{\mathrm{T}}(s)] = \int_a^m E[G(u)G^{\mathrm{T}}(u)] \, du, \tag{4.1.11}$$

where $m = \min\{t, s\}$; in particular,

$$E[\|x(t)\|^2] = \int_a^t E[\|G(u)\|^2] \, du, \tag{4.1.12}$$

and for all $c > 0$, $a \le t_1 \le t_2 \le b$,

$$P\left\{ \sup_{t_1 \le t \le t_2} \|x(t) - x(t_1)\| > c \right\} \le \frac{1}{c^2} \int_{t_1}^{t_2} E[\|G(s)\|^2] \, ds,$$

and

$$E\left[ \sup_{a \le t \le b} \|x(t) - x(a)\|^2 \right] \le 4 \int_a^b E[\|G(s)\|^2] \, ds.$$

(e)   $x(t)$ is a.s. sample continuous.

In the following, we discuss an approximate sum of a stochastic integral.
Let $f(t, x)$ and the column vectors of an $n \times m$ matrix $\sigma(t, x)$ be continuous functions defined on $[a, b] \times R^n$, and let $\Lambda$ be a parameter set. For each $\lambda \in \Lambda$ we have an a.s. stochastic process $x_\lambda \equiv x_\lambda(t)$ and a Wiener process $z_\lambda(t)$ such that the sub-$\sigma$-algebra $\mathscr{F}_{\lambda t}$ of $\mathscr{F}$ is nonanticipating relative to the $m$-dimensional Wiener process. Then by Example 4.1.2,

$$I_\lambda = \int_a^b f(s, x_\lambda(s)) \, ds + \int_a^b \sigma(s, x_\lambda(s)) \, dz_\lambda(s)$$

is well-defined. Let $\Delta I_\lambda$ be the approximate sum of $I_\lambda$ with respect to a partition $\Delta = [a - t_0 < t_1 < \cdots < t_k < \cdots < t_n = b]$, where $t_k = t_0 + k(b - a)/n$, $k = 0, 1, \ldots, n - 1$; that is,

$$\Delta I_\lambda = \sum_{k=0}^{n-1} f(t_k, x_\lambda(t_k))(t_{k+1} - t_k)$$

$$+ \sum_{k=0}^{n-1} \sigma(t_k, x_\lambda(t_k))[z_\lambda(t_{k+1}) - z_\lambda(t_k)].$$

For Example 4.1.2, it is clear that $\Delta I_\lambda \to I_\lambda$ in probability for each $\lambda$ as $\delta_n = \max\{(t_{k+1} - t_k)\} \to 0$; that is, for given $\varepsilon > 0$, there exists an $\alpha = \alpha(\varepsilon, \lambda) > 0$ such that $\delta_n < \alpha$ implies $P(\|I_\lambda - \Delta I_\lambda\| > \varepsilon) < \varepsilon$.

The following result provides sufficient conditions for the convergence of the approximate sum $\Delta I_\lambda$ to $I_\lambda$ uniformly in probability.

**Proposition 4.1.1.**   If $\{x_\lambda(t), \lambda \in \Lambda\}$ is totally $D$-bounded, then for given $\varepsilon > 0$, there exists an $\alpha = \alpha(\varepsilon) > 0$ independent of $x$ such that $\delta_n < \alpha$ implies $P(\|I_\lambda - \Delta I_\lambda\| > \varepsilon) < \varepsilon$ for all $\lambda \in \Lambda$.

*Proof.*   Since $\{x_\lambda(t)\}$ is totally $D$-bounded, for every $\varepsilon > 0$ there exists a compact subset $K_\varepsilon$ of $R^n$ such that

$$P(x_\lambda(t) \in K_\varepsilon) > 1 - \tfrac{1}{2}\varepsilon \qquad \text{for every} \quad \lambda \in \Lambda.$$

Since $f(t, x_\lambda(t))$ and $\sigma(t, x_\lambda(t))$ are continuous, they are uniformly continuous on $[a, b]$ whenever $x_\lambda(t) \in K_\varepsilon$. For $x_\lambda(t), y_\lambda(t) \in K_\varepsilon$, there exists an $\alpha = \alpha(\varepsilon)$ such that

$$|t - s| < \alpha \quad \text{for all} \quad t, s \in [a, b], \qquad \sup_{a \le t \le b} \|x_\lambda(t) - y_\lambda(t)\| < \alpha$$

implies

$$\|f(t, x_\lambda(t)) - f(s, y_\lambda(s))\| < \sqrt{\frac{\varepsilon^3}{2c}}$$

and

$$\|\sigma(t, x_\lambda(t)) - \sigma(s, y_\lambda(s))\| < \sqrt{\frac{\varepsilon^3}{2c}},$$

where $c = 2(b - a)^2 + 2(b - a)$; in particular,

$$|t - s| < \alpha \quad \text{for} \quad t, s \in [a, b] \qquad x_\lambda(t) \in K_\varepsilon$$

implies

$$\|f(t, x_\lambda(t)) - f(s, x_\lambda(s))\| < \sqrt{\frac{\varepsilon^3}{2c}}$$

and

$$\|\sigma(t, x_\lambda(t)) - \sigma(s, x_\lambda(s))\| < \sqrt{\frac{\varepsilon^3}{2c}}.$$

From this proof, Example 4.1.2, and the definition of the stochastic integral of a step function, $\Delta I_\lambda$ can be written as

$$\Delta I_\lambda = \int_a^b f''(s, x_\lambda(s))\, ds + \int_a^b \sigma''(s, x_\lambda(s))\, dz_\lambda(s),$$

where

$$f''(t, x_\lambda(t)) = f(t_k, x_\lambda(t_k)) I_{[t_k, t_{k+1}]}(t)$$

and

$$\sigma''(t, x_\lambda(t_k)) = \sigma(t_k, x_\lambda(t)) I_{[t_k, t_{k+1}]}(t)$$

for $k = 1, 2, \ldots, n - 1$, and so

$$I_\lambda - \Delta I_\lambda = \int_a^b [f(s, x_\lambda(s)) - f''(s, x_\lambda(s))]\, ds$$
$$+ \int_a^b [\sigma(s, x_\lambda(s)) - \sigma''(s, x_\lambda(s))]\, dz_\lambda(s).$$

Set $\eta = \sqrt{\varepsilon^3/2c}$. Define a truncated function $\psi$ relative to $\eta$ as follows:

$$\psi(\phi) = \begin{cases} \|\phi\| & \text{if } \|\phi\| \le \eta, \\ 0 & \text{if } \|\phi\| > 0. \end{cases}$$

Setting

$$\Delta J_\lambda = \int_a^b \psi(f(s, x_\lambda(s)) - f^n(s, x_\lambda(s))) \, ds$$
$$+ \int_a^b \psi(\sigma(s, x_\lambda(s)) - \sigma^n(s, x_\lambda(s))) \, dz_\lambda(s),$$

we can see that $\delta_n < \alpha(\varepsilon)$ implies

$$P(\Delta J_\lambda \ne I_\lambda - \Delta I_\lambda) \le P(x_\lambda(t) \notin K_\varepsilon) < \varepsilon/2.$$

On the other hand, $\delta_n < \alpha(\varepsilon)$ implies

$$E[\|\Delta J_\lambda\|^2] \le 2 \left[ E \left[ \left\| \int_a^b \psi(f(s, x_\lambda(s)) - f^n(s, x_\lambda(s))) \, ds \right\|^2 \right] \right.$$
$$\left. + \int_a^b E[\|\psi(\sigma(s, x_\lambda(s)) - \sigma^n(s, x_\lambda(s)))\|^2] \, ds \right]$$
$$\le 2 \left[ \int_a^b \eta \, ds \right]^2 + \left[ \int_a^b \eta^2 \, ds \right]$$
$$= 2[(b-a)^2 \eta^2 + \eta^2(b-a)]$$
$$= c\eta^2 = \varepsilon^3/2,$$

and hence by Chebyshev's inequality,

$$P(\|\Delta J_\lambda\| > \varepsilon) \le \frac{E[\|\Delta J_\lambda\|^2]}{\varepsilon^2} \le \frac{\varepsilon^3}{2\varepsilon^2} = \frac{\varepsilon}{2};$$

thus

$$P(\|I_\lambda - \Delta I_\lambda\| > \varepsilon) \le P(\Delta J_\lambda \ne I_\lambda - \Delta I_\lambda) + P(\|\Delta J_\lambda\| > \varepsilon) < \varepsilon.$$

This proves the desired statement.

We shall now introduce the notion of a stochastic differential. Let $x(t)$ be a stochastic indefinite integral defined by

$$x(t) = x(a) + \int_a^t f(s) \, ds + \int_a^t G(s) \, dz(s), \tag{4.1.13}$$

where $z(t)$ is an $m$-dimensional Wiener process, $G \in M_2[a, b]$, $x(a)$ is an $n$-dimensional random vector relative to $\mathscr{F}_a$, and $f$ is an $n$-dimensional non-anticipating random vector relative to a nonanticipating family $\{\mathscr{F}_t : t \in R_+\}$

of sub-$\sigma$-algebras of $\mathscr{F}$ and is sample Lebesgue-integrable on $[a, b]$, that is,

$$\int_a^b \|f(s, \omega)\| \, ds < \infty \quad \text{w.p. 1.}$$

We further note that the integrals $\int_a^t f(s) \, ds$ and $\int_a^t G(s) \, dz(s)$ are in the sense of Lebesgue and Itô, respectively.

**Definition 4.1.6.**  We shall say that the process $x(t)$ defined by Eq. (4.1.13) processes the stochastic differential $f(t) \, dt + G(t) \, dz(t)$, and we shall write

$$dx(t) = f(t) \, dt + G(t) \, dz(t). \tag{4.1.14}$$

**Remark 4.1.1.**  Let $z(t)$ be a scalar normalized Wiener process. By (4.1.8), the stochastic indefinite integral of $z(t)$ relative to $z(t)$ is given by

$$\int_a^t z(s) \, dz(s) = \frac{1}{2} [z^2(t) - z^2(a) - (t - a)].$$

Hence its stochastic differential form is equal to

$$d(z^2(t)) = 2z(t) \, dz(t) + dt. \tag{4.1.15}$$

On the other hand, by applying Taylor's theorem, we obtain

$$d(z^2(t)) = 2z(t) \, dz(t) + (dz(t))^2.$$

This suggests that in the case of the stochastic differential of $z^2(t)$, we must regard the first two terms as first-order terms and must replace $(dz(t))^2$ with $dt$.

We shall next derive a well-known differential formula known as Itô's formula.

**Theorem 4.1.1**  (Itô's formula).   Let $V \in C[R_+ \times R^n, R^N]$, and let $V_t$, $V_x$, and $V_{xx}$ exist and be continuous for $(t, x) \in R_+ \times R^n$, where $V_x$ is an $N \times n$ Jacobian matrix of $V(t, x)$ and $V_{xx}(t, x) =$ is an $n \times n$ Hessian matrix whose elements $(\partial^2/\partial x_i \, \partial x_j) V(t, x)$ are $N$-dimensional vectors. Suppose that an $n$-dimensional random function $x(t)$ has stochastic differential $dx(t) = f(t) \, dt + G(t) \, dz(t)$ and $x(a) = x_0$, where $z(t)$ is an $m$-dimensional normalized Wiener process, $G \in M_2[a, b]$, and $f$ is an $n$-dimensional nonanticipating random vector relative to the nonanticipating family $\{\mathscr{F}_t : t \in R_+\}$ of sub-$\sigma$-algebras of $\mathscr{F}$, which is sample Lebesgue-integrable on $[a, b]$. $x(a)$ is a random vector relative to $\mathscr{F}_a$. Then the process $m(t) = V(t, x(t))$ possesses a stochastic differential, and moreover

$$\begin{aligned} dm(t) = \, &(V_t(t, x(t)) + V_x(t, x(t))f(t) \\ &+ \tfrac{1}{2} \text{tr}(V_{xx}(t, x(t))G(t)G^T(t))) \, dt \\ &+ V_x(t, x(t))G(t) \, dx(t). \end{aligned} \tag{4.1.16}$$

*Proof.* Since the proof of the theorem is exactly similar to the proof of its scalar version ($n = m = N = 1$), we shall give only the proof of its scalar version.

Let us first prove the formula for the case in which $f(t)$ and $G(t)$ are independent of $t$. Then $x(t)$ is a process of the form $x(t) = x(a) + f(t - a) + G(z(t) - z(a))$, and hence the process $m(t)$ takes the form

$$m(t) = V(t, x(t)) = V(t, f(t - a) + G(z(t) - z(a)) + x(a))$$

with $m(a) = V(a, x_0)$.

Suppose that $t_0 = a < t_1 < \cdots < t_n = t \le b$. Then

$$m(t) - m(a) = \sum_{i=1}^{n} V(t_i, x(t_i)) - V(t_{i-1}, x(t_{i-1})). \qquad (4.1.17)$$

From the assumption on $V(t, x)$, Taylor's formula yields

$$
\begin{aligned}
V(t_i, x(t_i)) &- V(t_{i-1}, x(t_{i-1})) \\
&= V_t(t_{i-1} + \theta_i(t_i - t_{i-1}), x(t_{i-1}))(t_i - t_{i-1}) \\
&\quad + V_x(t_{i-1}, x(t_{i-1}))(x(t_i) - x(t_{i-1})) \\
&\quad + \tfrac{1}{2} V_{xx}(t_{i-1}, x(t_{i-1}) + \psi_i(x(t_i) - x(t_{i-1})))(x(t_i) - x(t_{i-1}))^2, \quad (4.1.18)
\end{aligned}
$$

where $\theta_i, \psi_i \in (0, 1)$ for all $i = 1, 2, \ldots, n$. From the sample continuity of the indefinite integral and the continuity of $V_t$ and $V_{xx}$, one can find an $\alpha$ that approaches zero w.p. 1 as $\delta_n \to 0$ and satisfies the inequalities

$$
\max_{1 \le i \le n} \left\{ \left| V_t(t_{i-1} + \theta_i(t_i - t_{i-1}), x(t_{i-1})) \right. \right.
$$

$$
\left. \left. - V_t(t_{i-1}, x(t_{i-1})) \right| \right\} < \alpha
$$

and

$$
\max_{1 \le i \le n} \left\{ \left| V_{xx}(t_{i-1}, x(t_{i-1}) + \psi_i(x(t_i) - x(t_{i-1}))) \right. \right.
$$

$$
\left. \left. - V_{xx}(t_{i-1}, x(t_{i-1})) \right| \right\} < \alpha,
$$

where $\delta_n = \max_{1 \le i \le n} \{\Delta t_i\}$, $\Delta t_i = t_i - t_{i-1}$. This, together with the facts

$$\sum_{i=1}^{n} (t_i - t_{i-1}) = (t - a),$$

$$\sum_{i=1}^{n} [x(t_i) - x(t_{i-1})]^2 \le 2f^2 \sum_{i=1}^{n} (\Delta t_i)^2 + 2G^2 \sum_{i=1}^{n} (z(t_i) - z(t_{i-1}))^2,$$

and

$$E\left[ \sum_{i=1}^{n} (z(t_i) - z(t_{i-1}))^2 \right] = t - a,$$

implies that

$$\sum_{i=1}^{n} V_t(t_{i-1} + \theta_i(t_i - t_{i-1}), x(t_{i-1}))(t_i - t_{i-1})$$

$$\xrightarrow{\text{a.s.}} \sum_{i=1}^{n} V_t(t_{i-1}, x(t_{i-1}))(t_i - t_{i-1}),$$

$$\sum_{i=1}^{n} V_{xx}(t_i, x(t_{i-1}) + \psi_i(x(t_i) - x(t_{i-1})))(x(t_i) - x(t_{i-1}))^2$$

$$\xrightarrow{\text{m.s.}} \sum_{i=1}^{n} V_{xx}(t_{i-1}, x(t_{i-1}))(x(t_i) - x(t_{i-1}))^2$$

whenever $\delta_n \to 0$. Now (4.1.18) and the fact

$$x(t_i) - x(t_{i-1}) = f(t_i - t_{i-1}) + G(z(t_i) - z(t_{i-1})),$$

reduce (4.1.17) to

$$m(t) - m(a) \xrightarrow{\text{i.p.}} \left[ \sum_{i=1}^{n} V_t(t_{i-1}, x(t_{i-1}))(t_i - t_{i-1}) \right.$$

$$+ \sum_{i=1}^{n} V_x(t_{i-1}, x(t_{i-1}))f(t_i - t_{i-1})$$

$$+ \sum_{i=1}^{n} V_x(t_{i-1}, x(t_{i-1}))G(z(t_i) - z(t_{i-1}))$$

$$+ \frac{1}{2} \sum_{i=1}^{n} V_{xx}(t_{i-1}, x(t_{i-1}))G^2(t_i - t_{i-1})$$

$$+ \frac{1}{2} \sum_{i=1}^{n} V_{xx}(t_{i-1}, x(t_{i-1}))[(z(t_i) - z(t_{i-1}))^2 - (t_i - t_{i-1})]G^2$$

$$+ \frac{1}{2} \sum_{i=1}^{n} V_{xx}(t_{i-1}, x(t_{i-1}))f^2(t_i - t_{i-1})^2$$

$$\left. + \sum_{i=1}^{n} V_{xx}(t_{i-1}, x(t_{i-1}))fG(z(t_i) - z(t_{i-1}))(t_i - t_{i-1}) \right].$$

This, together with the relations

$$f^2 \sum_{i=1}^{n} (t_i - t_{i-1})^2 \xrightarrow{\text{a.s.}} 0$$

and

$$G \sum_{i=1}^{n} (z(t_i) - z(t_{i-1}))(t_i - t_{i-1}) \xrightarrow{\text{i.p.}} 0$$

as $\delta_n \to 0$, the continuity of $V_t$, $V_x$; and $V_{xx}$, the almost-sure continuity of $z(t)$, and the definition of stochastic integral, yield

$$m(t) - m(t_0) = \int_{t_0}^{t} [V_t(s, x(s)) + V_x(s, x(s))f + \tfrac{1}{2}V_{xx}(s, x(s))G^2]\, ds$$

$$+ \int_{t_0}^{t} V_x(s, x(s))G\, dz(s), \qquad (4.1.19)$$

provided that

$$\sum_{i=1}^{n} V_{xx}(t_{i-1}, x(t_{i-1}))[(z(t_i) - z(t_{i-1}))^2 - (t_i - t_{i-1})]G^2 \xrightarrow{\text{i.p.}} 0 \quad (4.1.20)$$

as $\delta_n \to 0$. From (4.1.19), the hypotheses of the theorem, and the definition of stochastic differential, it is obvious that $m(t)$ possesses the stochastic differential, and moreover

$$dm(t) = (V_t(t, x(t)) + V_x(t, x(t))f$$
$$+ \tfrac{1}{2}V_{xx}(t, x(t))G^2)\, dt + V_x(t, x(t))G\, dz(t).$$

This proves the scalar version of (4.1.16), provided that (4.1.20) is valid. For proving (4.1.20), we set

$$S_n = \sum_{i=1}^{n} V_{xx}(t_{i-1}, x(t_{i-1}))[z(t_i) - z(t_{i-1}))^2 - (t_i - t_{i-1})]\, G^2.$$

To eliminate large values of $V_{xx}$, we truncate the function as follows: For positive number $K$, let us define

$$I_i^K(\omega) = \begin{cases} 1 & \text{if } |x(t_j)| \le K \quad \text{for all } j \le i, \\ 0 & \text{otherwise,} \end{cases}$$

$$\varepsilon_i = (z(t_i) - z(t_{i-1}))^2 - (t_i - t_{i-1}),$$

and

$$S_n^K = \sum_{i=1}^{n} V_{xx}(t_{i-1}, x(t_{i-1}))I_{i-1}^K G^2 \varepsilon_i.$$

We note that $E[\varepsilon_i] = 0$ for all $1 \le i \le n$, that $\varepsilon_i$ are independent of each other and of $V_{xx}(t_{i-1}, x(t_{i-1}))I_{i-1}^K G^2$, and that $E(\varepsilon_i^2) = 2(t_i - t_{i-1})^2$. It is obvious that $E[S_n^K] = 0$ and

$$E[(S_n^K)^2] = \sum_{i=1}^{n} E[(V_{xx}(t_{i-1}, x(t_{i-1}))I_{i-1}^K)^2 G^2 E(\varepsilon_i^2)]$$

$$\le 2 \max_{\substack{t_0 \le s \le t \\ |x| \le K}} \{|V_{xx}(t, x)|G^2\} \sum_{i=1}^{n} (t_i - t_{i-1})^2,$$

which tends to zero as $\delta_n \to 0$. Therefore, for every fixed $K > 0$, we have

$$S_n^K \xrightarrow{\text{m.s.}} 0 \qquad \text{implies} \qquad S_n^K \xrightarrow{\text{i.p.}} 0.$$

The error resulting from the trucation is

$$P\{S_n \neq S_n^K\} = P\left\{ \max_{t_0 \leq s \leq t} |x(s)| > K \right\}. \tag{4.1.21}$$

We observe that

$$\max_{t_0 \leq s \leq t} |x(s)| = \max_{t_0 \leq s \leq t} |x(t_0) + f(s - t_0) + G(z(s) - z(t_0))|$$

$$\leq |x_0| + |f|(t - a) + |G| \max_{t_0 \leq s \leq t} |z(s) - z(a)|,$$

which is an a.s. finite random variable, so that the right-hand quantity in (4.1.21) can be made arbitrarily small by choosing $K$ sufficiently large. Hence given $\eta > 0$ and $\delta > 0$, there exists an integer $N_0$ such that

$$P\{|S_n| > \delta\} \leq P\{|S_n^K| > \delta\} + P\{S_n \neq S_n^K\} < \eta, \qquad n \geq N_0.$$

This implies that $S_n$ converges to zero in probability and establishes the validity of (4.1.20). The proof is complete.

The following example gives an indication of the possible uses of Itô's formula.

**Example 4.1.3.** Consider a scalar process given by

$$dx = f(t) dt + G(t) dz(t), \qquad x(a) = x_0,$$

where $f, G \in M_2[a, b]$. Let us compute the stochastic differential of $m(t) = e^{x(t)}$. Applying Itô's formula to $e^x$, we get

$$dm = e^x f(t) dt + \tfrac{1}{2} e^x G^2(t) dt + e^x G(t) dz(t)$$

$$= m[(f(t) + \tfrac{1}{2} G^2(t)) dt + G(t) dz(t)],$$

$$m(a) = e^{x_0}.$$

This is a linear stochastic differential equation for the process $m(t)$. It is clear from the foregoing derivation that

$$m(t) \equiv m(a) \exp\left[ \int_a^t f(s) ds + \int_a^t G(s) dz(s) \right], \qquad t \in [a, b].$$

We can also solve the above linear equation directly by using Itô's formula. For example, set $v = \log m$, and apply Itô's formula for $\log m$ to get

$$dv = \frac{1}{m} [(f(t) + \tfrac{1}{2} G^2(t)) m dt + G(t) m dz(t)] - \frac{1}{2} \frac{1}{m^2} [G^2(t) m^2 dt]$$

$$= f(t) dt + G(t) dz(t).$$

Thus

$$v(t) = v(a) + \int_a^t f(s)\,ds + \int_a^t G(s)\,dz(s),$$

or equivalently,

$$m(t) = m(a)\exp\left[\int_a^t f(s)\,ds + \int_a^t G(s)\,dz(s)\right], \qquad t \in [a, b], \quad (4.1.22)$$

which is a solution of the linear equation considered.

Finally we shall present another simple application of Itô's formula that will be employed subsequently.

**Lemma 4.1.2.**  Assume that $G \in M_2[a, b]$ and

$$\int_a^b E(\|G(s)\|^{2k})\,ds < \infty,$$

where $k$ is some positive integer. Then

$$E\left[\left\|\int_a^b G(t)\,dz(t)\right\|^{2k}\right] \le (k(2k-1))^k(b-a)^{k-1}\int_a^b E[\|G(t)\|^{2k}]\,dt. \quad (4.1.23)$$

*Proof.*  Applying Itô's formula to the function $V(x) = \|x\|^{2k}$ relative to the stochastic differential $dx(t) = G(t)\,dz(t)$ with $x(a) = 0$, we obtain

$$\begin{aligned}
d(\|x(t)\|^{2k}) = {} & 2k\|x(t)\|^{2k-2}x^T(t)G(t)\,dz \\
& + k\|x(t)\|^{2k-2}\,\mathrm{tr}(G(t)G^T(t))\,dt \\
& + 2k(k-1)\|x(t)\|^{2k-4}\|x^T(t)G(t)\|^2\,dt,
\end{aligned}$$

which is equivalent to

$$\begin{aligned}
\|x(b)\|^{2k} = {} & \int_a^b 2k\|x(s)\|^{2k-2}x^T(s)G(s)\,dz(s) \\
& + \int_a^b k\|x(s)\|^{2k-2}\,\mathrm{tr}(G(s)G^T(s))\,ds \\
& + \int_a^b 2k\|x(s)\|^{2k-2}x^T(s)G(s)\,dz(s)
\end{aligned}$$

Since $x(b) = \int_a^b G(s)\,dz(s)$, we have

$$\begin{aligned}
\left\|\int_a^t G(s)\,dz(s)\right\|^{2k} = {} & \int_a^t 2k\|x(s)\|^{2k-2}x^T(s)G(s)\,dz(s) \\
& + \int_a^t k\|x(s)\|^{2k-2}\,\mathrm{tr}(G(s)G^T(s))\,ds \\
& + \int_a^b 2k(k-1)\|x(s)\|^{2k-4}\|x^T(s)G(s)\|^2\,ds. \quad (4.1.24)
\end{aligned}$$

From the nature of $G$, one can construct a sequence of step functions $G_m \in M_2[a, b]$ such that $\int_a^b E[\|G(t) - G_n(t)\|^{2k}] \, ds \to 0$, that is, the class of step functions that are dense in $L^{2k}(\Omega)$. By the definition of the stochastic integral, it is obvious that

$$\int_a^b E\left[ \left\| \int_a^s G_n(s) \, dz(s) \right\|^{2k-2} \|G_n(s)\|^2 \right] ds < \infty;$$

thus for the sequence $G_n$, relation (4.1.24) reduces to

$$E\left[ \left\| \int_a^t G_n(s) \, dz(s) \right\|^{2k} \right] = E\left[ k \int_a^t \|x(s)\|^{2k-2} \operatorname{tr}(G_n(s)G_n^{\mathrm{T}}(s)) \, ds \right.$$

$$\left. + 2k(k-1) \int_a^t \|x(s)\|^{2k-4} \|x^{\mathrm{T}}(s)G_n(s)\|^2 \, ds \right]. \quad (4.1.25)$$

By using the properties of the norm, we get

$$E\left[ \left\| \int_a^b G_n(s) \, dz(s) \right\|^{2k} \right] \leq E\left[ k \int_a^t [\|x(s)\|^{2k-2} \|G_n(s)\|^2] \, ds \right.$$

$$\left. + 2k(k-1) \int_a^t [\|x(s)\|^{2k-2} \|G_n(s)\|^2] \, ds \right]$$

$$\leq E\left[ k(2k-1) \int_a^t [\|x(s)\|^{2k-2} \|G_n(s)\|^2] \, ds \right]$$

$$\leq E\left[ k(2k-1) \int_a^t \left[ \left\| \int_a^s G_n(u) \, dz(u) \right\|^{2k-2} \|G_n(s)\|^2 \right] ds \right].$$

This, together with the independence of $\|\int_a^s G_n(u) \, dz(u)\|^{2k-2}$ and $\|G_n(s)\|^2$, shows that

$$E\left[ \left\| \int_a^t G_n(s) \, dz(s) \right\|^{2k} \right]$$

$$\leq k(2k-1) \int_a^t \left[ E\left[ \left\| \int_a^s G_n(u) \, dz(u) \right\|^{2k-2} \right] \right] [E[\|G_n(s)\|^2]] \, ds.$$

Applying Hölder's inequality, we have

$$E\left[ \left\| \int_a^b G_n(s) \, dz(s) \right\|^{2k} \right] \leq k(2k-1) \left[ \int_a^b E\left[ \left\| \int_a^s G_n(u) \, dz(u) \right\|^{2k} \right] ds \right]^{(2k-2)/2k}$$

$$\times \left[ \int_a^b E[\|G_n(s)\|^{2k}] \, ds \right]^{1/k}.$$

From (4.1.25), it is evident that $E[\|\int_a^t G_n(s)\,dz(s)\|^{2k}]$ is increasing in $t$. Hence

$$E\left[\left\|\int_a^b G_n(s)\,dz(s)\right\|^{2k}\right] \leq k(2k-1)\left[\int_a^b E\left[\left\|\int_a^b G_n(u)\,dz(u)\right\|^{2k}\right]ds\right]^{(k-1)/k}$$

$$\times \left[\int_a^b E[\|G_n(s)\|^{2k}]\,ds\right]^{1/k}.$$

Taking the $k$th power and simplifying both sides of the inequality, we get

$$E\left[\left\|\int_a^b G_n(s)\,dz(s)\right\|^{2k}\right] \leq (k(2k-1))^k(b-a)^{k-1}\int_a^b E[\|G_n(s)\|^{2k}]\,ds.$$

Now taking the limit as $n \to \infty$, we have the desired inequality and the proof of the Lemma is complete.

Finally, we shall formulate Itô-type stochastic differential systems. In order to do this, we shall first consider a special class of Markov processes with a.s. sample continuous functions.

**Definition 4.1.7.**   A Markov process $x(t)$ for $t \in I$, with values in $R^n$ and a.s. sample continuous, is called a diffusion process if its transition probability $P(s, x, t, B)$ satisfies the following three Hinčin conditions for every $s \in I$, $x \in R^n$, and $\varepsilon > 0$:

$$\int_{\|y-x\|>\varepsilon} P(s,x,t,dy) = o(t-s), \tag{4.1.26}$$

$$\int_{\|y-x\|\leq\varepsilon} (y-x)P(s,x,t,dy) = f(s,x)(t-s) + o(t-s), \tag{4.1.27}$$

and

$$\int_{\|y-x\|\leq\varepsilon} (y-x)(y-x)^{\mathsf{T}}P(s,x,t,dy) = b(s,x)(t-s) + o(t-s), \tag{4.1.28}$$

where $f(s,x)$ is an $R^n$-valued function and $b(s,x)$ is an $n \times n$ matrix-valued function. The functions $f$ and $b$ are called the coefficients of the diffusion process. In particular, $f$ is called the drift vector, and $b$ is called the diffusion matrix. $b$ is a symmetric and nonnegative definite matrix.

In the homogeneous case, $P(s, x, t, B)$ depends only on $(t-s)$. We set $P(t, x, B) = P(0, x, t, B)$; thus $P(s, x, t, B) = P(t-s, x, A)$, and the Hinčin conditions reduce to

$$\int_{\|y-x\|>\varepsilon} P(t,x,dy) = o(t),$$

$$\int_{\|y-x\|\leq\varepsilon} (y-x)P(t,x,dy) = f(x)t + o(t),$$

and

$$\int_{||y-x|| \le \varepsilon} (y - x)(y - x)^{\mathrm{T}} P(t, x, dy) = b(x)t + o(t),$$

for any $x \in R^n$, $\varepsilon > 0$.

**Remark 4.1.2.**   For the process $x(t)$ to be a diffusion process it is sufficient that the Kolmogorov condition implies the Hiňcin conditions, that is, that for some $\delta > 0$,

$$\int_{R^n} ||y - x||^{2+\delta} P(s, x, t, dy) = o(t - s)$$

implies relations (4.1.26)–(4.1.28). In fact, for $k = 0, 1, 2$ we have

$$\int_{||y-x|| > \varepsilon} ||y - x||^k P(s, x, t, dy)$$

$$\le \varepsilon^{-2-\delta+k} \int_{R^n} ||y - x||^{2+\delta} P(s, x, t, dy) = o(t - s)$$

for all $\varepsilon > 0$.

Let us discuss the meaning of relations (4.1.26)–(4.1.28). Relation (4.1.26) means that large changes in $x(t)$ over a short period of time are improbable, that is,

$$P(||x(t) - x(s)|| \le \varepsilon \,|\, x(s) = x) = 1 - o(t - s).$$

We replace the truncated moments in (4.1.27) and (4.1.28) with the usual ones. Then the first moment of $x(t) - x(s)$ under the condition $x(s) = x$ as $t \to s^+$ is

$$E[(x(t) - x(s)) \,|\, x(s) = x] = \int_{||y-x|| \le \varepsilon} (y - x) P(s, x, t, dy)$$

$$+ \int_{||y-x|| > \varepsilon} (y - x) P(s, x, t, dy)$$

$$= f(s, x)(t - s) + o(t - s),$$

and

$$E[(x(t) - x(s))(x(t) - x(s))^{\mathrm{T}} \,|\, x(s) = x]$$

$$= \int_{||y-x|| \le \varepsilon} (y - x)(y - x)^{\mathrm{T}} P(s, x, t, dy)$$

$$+ \int_{||y-x|| > \varepsilon} (y - x)(y - x)^{\mathrm{T}} P(s, x, t, dy)$$

$$= b(s, x)(t - s) + o(t - s).$$

Therefore,

$$\mathrm{cov}\,[(x(t) - x(s)) \,|\, x(s) = x] = b(s, x)(t - s) + o(t - s),$$

where $\text{cov}\left[(x(t) - x(s))\big| x(s) = x\right]$ is the covariance matrix of $x(t) - x(s)$ with respect to the probability $P(s, x, t, \cdot)$. Consequently, $f(s, x)$ is the mean velocity vector of the random motion described by $x(t)$ under the assumption that $x(s) = x$, whereas $b(s, x)$ is a measure of the local magnitude of the fluctuation of $x(t) - x(s)$ about the mean value. If we neglect the last term $o(t - s)$, we can write

$$x(t) - x(s) \approx f(s, x(s))(t - s) + \sigma(s, x, (s))[w(t) - w(s)],$$

where

$$E\left[(w(t) - w(s))\big| x(s) = x\right] = 0,$$
$$\text{cov}\left[(w(t) - w(s))\big| x(s) = x\right] = I(t - s),$$

$\sigma(s, x)$ is any $n \times m$ matrix with the property $\sigma\sigma^T = b$, and $I$ is the identity matrix. The shift to differentials yields

$$dx(t) = f(t, x(t)) \, dt + \sigma(t, x(t)) \, dw(t). \tag{4.1.29}$$

This is the Itô-type stochastic differential systems or diffusion differential systems.

## 4.2. EXISTENCE AND UNIQUENESS

Let $\Omega = (\Omega, \mathscr{F}, P)$ be a complete probability space, and let $R[\Omega, R^n]$ be the system of all $R^n$-valued random variables. Let $J = [t_0, t + a]$, $t_0 \geq 0$, $a > 0$.
We consider the stochastic differential system of Itô-type

$$dx = f(t, x) \, dt + \sigma(t, x) \, dz, \qquad x(t_0) = x_0, \tag{4.2.1}$$

where $dx$ is a stochastic differential of $x$,

$$x, f \in C[J \times R^n, R^n], \qquad \sigma \in C[J \times R^n, R^{nm}],$$

and $z(t) \in R[\Omega, R^m]$ is a normalized Wiener process.
Let us now define a solution process for the stochastic system (4.2.1).

**Definition 4.2.1.** A random process $x(t)$ is called a solution process or a solution of (4.2.1) on $J$ if it satisfies the following conditions:

(1)  $x(t)$ is $\mathscr{F}_t$-measurable, that is, nonanticipating, where for $t \in J$, $\mathscr{F}_t$ is the smallest sub-$\sigma$-algebra generated by $x_0$ and $z(t)$;
(2)  $x(t)$ is sample continuous and verifies

$$\int_{t_0}^{t} \left[\|f(s, x(s))\| + \|\sigma(s, x(s))\|^2\right] ds < \infty \quad \text{w.p. 1};$$

(3)  $x(t)$ satisfies Eq. (4.2.1) for $t \in J$ w.p. 1;

We have intentionally chosen to consider the deterministic functions $f$ and $\sigma$ in the Itô-type equation (4.2.1). However, one could utilize random functions $f$ and $\sigma$ as long as they were nonanticipating relative to $\mathscr{F}_t$. This change does not in any way create significant problems in our discussion.

We shall prove the following basic existence theorem of Peano-type.

**Theorem 4.2.1.**   Assume that

(H$_1$)   there exist positive numbers $L$ and $M$ such that

$$\|f(t, x)\| + \|\sigma(t, x)\| \le L + M\|x\|$$

for $(t, x) \in J \times R^n$;

(H$_2$)   $x(t_0)$ is independent of $z(t)$ and $E[\|x(t_0)\|^4] \le c_1$ for some constant $c_1 > 0$.

Then the stochastic system (4.2.1) possesses at least one solution $x(t) = x(t, t_0, x_0)$ on $J$.

*Proof.*   We define a sequence of random functions $\{x_n(t)\}$ on $J$ by setting

$$x_n(t) = \begin{cases} x_0 & \text{for} \quad t_0 \le t \le t_0 + a/n \\ x_0 + \displaystyle\int_{t_0}^{t-a/n} f(s, x_n(s))\, ds + \int_{t_0}^{t-a/n} \sigma(s, x_n(s))\, dz(s) \end{cases} \tag{4.2.2}$$

for $t_0 + ka/n \le t \le t_0 + (k + 1)a/n$, $k = 1, 2, \ldots, n - 1$. The first part defines $x_n(t)$ on $[t_0, t_0 + a/n]$, the second part defines $x_n(t)$, first on $[t_0 + a/n, t_0 + 2a/n]$, then on the interval $[t_0 + 2a/n, t_0 + 3a/n]$, and so on. In fact,

$$x_n(t) = x_n\left(t_0 + \frac{ka}{n}\right) + \int_{t_0 + ka/n}^{t} f\left(s - \frac{ka}{n}, x_n\left(s - \frac{ka}{n}\right)\right) ds$$

$$+ \int_{t_0 + ka/n}^{t} \sigma\left(s - \frac{ka}{n}, x_n\left(s - \frac{ka}{n}\right)\right) dz(s) \tag{4.2.3}$$

for $t \in [t_0 + ka/n, t_0 + (k + 1)a/n]$, $k = 1, 2, \ldots, n - 1$.

By the continuity of $f$ and $\sigma$ and (H$_2$), $f(t, x_n(t))$ and $\sigma(t, x_n(t))$ are nonanticipating random functions relative to the family $\mathscr{F}_t$. Consequently, $\sigma(t, x_n(t)) \in M_2[t_0, t_0 + a]$ and $f(t, x_n(t))$ is integrable. Thus the sequence $\{x_n(t)\}$ satisfies the properties of a stochastic indefinite integral.

We shall first show that there exists a positive number $\alpha$ independent of $n$, such that

$$E[\|x_n(t)\|^4] \le \alpha \qquad \text{for} \quad t \in J. \tag{4.2.4}$$

We note that because of $(H_2)$,

$$E[\|x_n(t)\|^4] \leq c_1 \qquad \text{for all} \quad t \in [t_0, t_0 + a/n]. \qquad (4.2.5)$$

Using the inequalities $(a + b + c)^4 \leq 3^3(a^4 + b^4 + c^4)$, $2a^2b^2 \leq a^4 + b^4$, $4a^3b \leq 3a^4 + b^4$, Hölder's inequality, Lemma 4.1.2, $(H_1)$, and (4.2.3), we obtain successively

$$E[\|x_n(t)\|^4] \leq 3^3 \left[ E\left[\left\|x_n\left(t + \frac{ka}{n}\right)\right\|^4\right] \right.$$

$$+ \frac{a^3}{n^3} \int_{t_0 + ka/n}^{t} E\left[\left\|f\left(s - \frac{ka}{n}, x_n\left(s - \frac{ka}{n}\right)\right)\right\|^4\right] ds$$

$$+ \left. \frac{6a}{n} \int_{t_0 + ka/n}^{t} E\left[\left\|\sigma\left(s - \frac{ka}{n}, x_n\left(s - \frac{ka}{n}\right)\right)\right\|^4\right] ds \right]$$

$$\leq 3^3 \left[ E\left[\left\|x\left(t_0 + \frac{ka}{n}\right)\right\|^4\right] \right.$$

$$+ \left. \left(\frac{a^3}{n^3} + \frac{6a}{n}\right) \int_{t_0 + ka/n}^{t} E\left[\left(L + ME\left[\left\|x_n\left(s - \frac{ka}{n}\right)\right\|\right]\right)^4\right] ds \right]$$

$$\leq 3^3 \left[ E\left[\left\|x_n\left(t_0 + \frac{ka}{n}\right)\right\|^4\right] \right.$$

$$+ \left. 8(a^3 + 6a) \int_{t_0 + ka/n}^{t} \left(L^4 + M^4 E\left[\left\|x_n\left(s - \frac{ka}{n}\right)\right\|^4\right]\right) ds \right]$$

for $t \in [t_0 + ka/n, t_0 + (k + 1)a/n]$. Starting now from (4.2.5), it is easy to see by induction that $E[\|x_n(t)\|^4] < \infty, t \in J$.

By Itô's formula, we get

$$\|x_n(t)\|^4 = \left\|x_n\left(t_0 + \frac{a}{n}\right)\right\|^4 + \int_{t_0 + a/n}^{t} \left[ 4\|x_n(s)\|^2 x_n^T(s) f\left(s - \frac{a}{n}, x_n\left(s - \frac{a}{n}\right)\right) \right.$$

$$+ \left. \text{tr}\left(b\left(s - \frac{a}{n}, x_n\left(s - \frac{a}{n}\right)\right) \frac{\partial^2}{\partial x \, \partial x} \|x_n(s)\|^4\right) \right] ds$$

$$+ 4 \int_{t_0}^{t} \|x_n(s)\|^2 x_n^T(s) \sigma\left(s - \frac{a}{n}, x_n\left(s - \frac{a}{n}\right)\right) dz(s) \qquad (4.2.6)$$

for $t \in [t_0 + a/n, t_0 + a]$, and

$$\|x_n(t)\|^4 = \|x_0\|^4 \qquad \text{for} \quad t \in [t_0, t_0 + a/n], \qquad (4.2.7)$$

where $b(t, x) = \sigma(t, x)\sigma^T(t, x)$. Since $E[\|x_n(t)\|^4] < \infty$, it follows by $(H_1)$ that $E[\|x_n(s)\|^3\|\sigma(s - a/n, x_n(s - a/n))\|] < \infty$. Thus (4.2.6) reduces to

$$E[\|x_n(t)\|^4] \leq E\left[\left\|x\left(t_0 + \frac{a}{n}\right)\right\|^4\right]$$

$$+ E\left[\int_{t_0 + a/n}^{t}\left[\|x_n(s)\|^3 4\left(L + M\left\|x_n\left(s - \frac{a}{n}\right)\right\|\right)\right.\right.$$

$$\left.\left. + 6\|x_n(s)\|^2\left(L + M\left\|x_n\left(s - \frac{a}{n}\right)\right\|\right)^2\right]ds\right],$$

which implies

$$m(t) \leq \alpha_1 + \alpha_2 \int_{t_0}^{t} m(s)\,ds$$

in view of relations (4.2.5) and (4.2.7) and the inequalities $4a^3b \leq 3a^4 + b^4$, $2a^2b^2 \leq a^4 + b^4$, where $\alpha_1 = c_1 + (L + 3L^2 + 3LM)a$, $\alpha_2 = 3L + 4M + 3L^2 + 6M^2 + 9LM$, and $m(t) = \sup_{-a/n \leq \theta \leq 0} E[\|x_n(t + \theta)\|^4]$. This, together with the scalar version of Theorem 1.7.2 with $g(t, m(t)) \equiv \alpha_2 m(t)$ and $m(t_0) = \alpha_1$, yields

$$m(t) \leq \alpha_1 \exp[\alpha_2 a], \qquad t \in J,$$

which proves relation (4.2.4) with $\alpha = \alpha_1 \exp[\alpha_2 a]$.

We shall now prove that there exists a positive number $\beta$ independent of $n$ such that

$$E[\|x_n(t) - x_n(s)\|^4] \leq \beta|t - x|^{3/2} \tag{4.2.8}$$

for all $t, s \in J$.

For this purpose, we apply Itô's formula to the expression $\|x_n(t) - x_n(s)\|^4$ and obtain

$$E[\|x_n(t) - x_n(s)\|^4]$$

$$= E\left[\int_{s}^{t}\left[\|x_n(u) - x_n(s)\|^2 4(x_n(u) - x_n(s))^T f\left(u - \frac{a}{n}, x_n\left(u - \frac{a}{n}\right)\right)\right.\right.$$

$$\left.\left. + \mathrm{tr}\left(b\left(u - \frac{a}{n}, x_n\left(u - \frac{a}{n}\right)\right)\frac{\partial^2}{\partial x\,\partial x}\|x_n(u) - x_n(s)\|^4\right)\right]du\right]$$

$$\leq \int_{s}^{t} E\left[4\|x_n(u) - x_n(s)\|^3\left\|f\left(u - \frac{a}{n}, x_n\left(u - \frac{a}{n}\right)\right)\right\|\right.$$

$$\left. + 6\|x_n(u) - x_n(s)\|^2\left\|\sigma\left(u - \frac{a}{n}, x_n\left(u - \frac{a}{n}\right)\right)\right\|^2\right]du.$$

Applying Hölder's inequality, we then have

$$E[\|x_n(t) - x_n(s)\|^4] \leq 2 \int_s^t \left[ \left( E\left[ \|x_n(u) - x_n(s)\|^4 \right] \right)^{1/2} \right.$$

$$\times \left( E\left[ \left( 2\|x_n(u) - x_n(s)\| \left\| f\left(u - \frac{a}{n}, x_n\left(u - \frac{a}{n}\right)\right) \right\| \right. \right. \right.$$

$$\left. \left. \left. + 3 \left\| \sigma\left(u - \frac{a}{n}, x_n\left(u - \frac{a}{n}\right)\right) \right\|^2 \right)^2 \right] \right)^{1/2} \right] du,$$

which gives further

$$E\left[ \|x_n(t) - x_n(s)\|^4 \right]$$

$$\leq 8\left[ \int_s^t (E[\|x_n(u) - x_n(s)\|^4])^{1/2} \left( E\left[ \|x_n(u) - x_n(s)\|^4 \right. \right. \right.$$

$$\left. \left. + \left\| f\left(u - \frac{a}{n}, x_n\left(u - \frac{a}{n}\right)\right) \right\|^4 + \left\| \sigma\left(u - \frac{a}{n}, x_n\left(u - \frac{a}{n}\right)\right) \right\|^4 \right] \right)^{1/2} du \right]$$

$$\leq 8\left[ \int_s^t (E[\|x_n(u) - x_n(s)\|^4])^{1/2} \left( 8\left( 2L^4 + 2M^4 E\left[ \left\| x_n\left(u - \frac{a}{n}\right) \right\|^4 \right] \right. \right. \right.$$

$$\left. \left. + E[\|x_n(u)\|^4] + E[\|x_n(s)\|^4] \right) \right)^{1/2} du \right].$$

Here we have utilized the inequalities $2a^2b^2 \leq a^4 + b^4$ and $4abc^2 \leq a^4 + b^4 + 2c^4$, the identity $(2ab + 3c^2)^2 = 4a^2b^2 + 12abc^2 + 9c^4$, and hypothesis $(H_1)$, together with suitable majorization to the nearest perfect cube number. It now follows from (4.2.4) that

$$E[\|x_n(u) - x_n(s)\|^4] \leq M_1 \int_s^t (E[\|x_n(u) - x_n(s)\|^4])^{1/2} du, \qquad (4.2.9)$$

where $M_1 = 32(\alpha + L^4 + M^4\alpha)^{1/2}$.

On the other hand, we have

$$E[\|x_n(t) - x_n(s)\|^4] \leq 8(E[\|x_n(t)\|^4] + E[\|x_n(s)\|^4]),$$

which because of (4.2.4), shows that $E[\|x_n(t) - x_n(s)\|^4] \leq 16\alpha$. It now follows from (4.2.9) that

$$E[\|x_n(t) - x_n(s)\|^4] \leq \tfrac{3}{2}\beta(t - s),$$

where $\beta = \tfrac{8}{3}M_1(\alpha)^{1/2}$. This, together with (4.2.9), establishes the inequality

$$E[\|x_n(t) - x_n(s)\|^4] \leq \beta(t - s)^{3/2} \qquad \text{for} \quad t_0 \leq s \leq t \leq t_0 + \alpha.$$

For $t_0 \leq t \leq s \leq t_0 + a$, the above inequality can be proved similarly, which proves inequality (4.2.8).

We recall that the collection $\{x_n(t)\}$ consists of continuous random processes defined on $J$ into $R[\Omega, R^n]$. In view of relations (4.2.4) and (4.2.8) and Lemma 2.1.1, we conclude that the collection is totally $D$-bounded. Using Prohorov's theorem and recalling the fact that the direct product of compact sets and the projection of a compact set are also compact, we can easily see that $\{(x_n(t), z_n(t), x_{n0})\}$ is totally bounded, where $z_n(t) \equiv z(t)$ and $x_{n0} \equiv x(t_0)$. Therefore, one can find a $D$-Cauchy subsequence $\{(x_{n_r}(t), z_{n_r}(t), x_{n_r0})\}$ of $\{(x_n(t), z_n(t), x_{n0})\}$. By Skorokhod's theorem, we can construct a sequence $\{(y_{n_r}(t), w_{n_r}(t), y_{n_r0})\}$ and a random function $(y(t), w(t), y_0)$ on a certain probability measure space such that

$$D((x_{n_r}(t), z_{n_r}(t), x_{n_r0}), (y_{n_r}(t), w_{n_r}(t), y_{n_r0})) = 0 \qquad (4.2.10)$$

for $n_1, n_2, n_3, \ldots$ and

$$P((y_{n_r}(t), w_{n_r}(t), y_{n_r0}) \to (y(t), w(t), y_0)) = 1 \qquad (4.2.11)$$

as $r \to \infty$, where the convergence is understood in the sense of the sup norm. For the sake of simplicity, hereafter we shall denote the subscript $n_r$ by just $r$. From $(H_2)$, (4.2.2), and (4.2.10), it follows that $w_r(t)$ and $y_{r0}$ are independent. This, together with (4.2.11), implies that $w(t)$ and $y_0$ are independent. Since $D((w_r(t), y_{r0}), z(t), x_0)) = 0$ for $r = 1, 2, \ldots$ and $P((w_r(t), y_{r0}) \to (w(t), y_0)) = 1$ as $r \to \infty$, we have

$$D((w(t), y_0), (z(t), x_0)) = 0, \qquad (4.2.12)$$

which implies that $(w(t), y_0)$ and $(z(t), x_0)$ have the same probability law. Hence the sub-$\sigma$-algebra generated by $w(t)$ and $y_0$ can be identified as $\mathscr{F}_t$. Because of (4.2.10) and (4.2.11), it is obvious that $y_r(t)$ and $y(t)$ possess all the properties of $x_n(t)$ as defined in (4.2.2). Furthermore, $\{y_r(t)\}$ is a $D$-Cauchy sequence. Finally, we shall prove that

$$y(t) = y_0 + \int_{t_0}^t f(s, y(s)) \, ds + \int_{t_0}^t \sigma(s, y(s)) \, dz(s). \qquad (4.2.13)$$

By following the definition of $x_n(t)$ in (4.2.2), one can construct

$$y_r^n(t) = y_{r0}, \qquad y^n(t) = y_0 \qquad \text{for} \quad t_0 \le t \le t_0 + a/n$$

and

$$y_r^n(t) = y_{r0} + \int_{t_0}^{t-a/n} f(s, y_r^n(s)) \, ds + \int_{t_0}^{t-a/n} \sigma(s, y_r^n(s)) \, dw_r(s),$$

$$y^n(t) = y_0 + \int_{t_0}^{t-a/n} f(s, y^n(s)) \, ds + \int_{t_0}^{t-a/n} \sigma(s, y^n(s)) \, dw(s)$$

$$\text{for} \quad t_0 + ka/n \le t \le t_0 + (k+1)a/n, \quad k = 1, 2, \ldots, n-1.$$

It is obvious that $\{y^n(t)\}$ and $\{y^n_r(t)\}$ for $r = 1, 2, \ldots$ possess all the properties of $\{x_n(t)\}$ in (4.2.2). This, together with the continuity of $f$ and $\sigma$, the fact that $\{y_r(t)\}$ is a $D$-Cauchy sequence, relation (4.2.11), Proposition 4.1.1, and Example 4.1.2, yields for any $r = 1, 2, \ldots$ and given $\varepsilon > 0$,

$$P(\|I^n_r(t) - I_r(t)\| > \varepsilon) < \varepsilon \qquad \text{and} \qquad P(\|I^n(t) - I(t)\| > \varepsilon) < \varepsilon \qquad (4.2.14)$$

for all $n \geq N(\varepsilon)$, where

$$I_r(t) = \int_{t_0}^t f(s, y_r(s)) \, ds + \int_{t_0}^t \sigma(s, y_r(s)) \, dw_r(s),$$

$$I(t) = \int_{t_0}^t f(s, y(s)) \, ds + \int_{t_0}^t \sigma(s, y(s)) \, dw(s),$$

$$I^n_r(t) = \int_{t_0}^{t-u/n} f(s, y^n_r(s)) \, ds + \int_{t_0}^{t-a/n} \sigma(s, y^n_r(s)) \, dw_r(s),$$

and

$$I^n(t) = \int_{t_0}^{t-a/n} f(s, y^n(s)) \, ds + \int_{t_0}^{t-a/n} \sigma(s, y^n(s)) \, dw(s).$$

Set

$$I^r_r(t) = \int_{t_0}^{t-a/r} f(s, y^r_r(s)) \, ds + \int_{t_0}^{t-a/r} \sigma(s, y^r_r(s)) \, d\omega_r(s).$$

Again, by the continuity of $f$ and $\sigma$ and (4.2.11), it is obvious that

$$P(I^n_r(t) \to I^n(t)) = 1 \qquad \text{as} \quad r \to \infty.$$

This implies that for any $\varepsilon > 0$, there exists a positive number $r$ such that $r \geq r_0 = r_0(\varepsilon)$,

$$\|f(s, y^n_r(s)) - f(s, y^n(s))\| < \sqrt{\varepsilon^3/4a^2},$$

and

$$\|\sigma(s, y^n_r(s)) - \sigma(s, y^n(s))\| < \sqrt{\varepsilon^3/4a};$$

hence

$$E[\|I^n_r(t) - I^n(t)\|^2] \leq 2E\left[\int_{t_0}^{t_0+a} \|f(s, y^n_r(s)) - f(s, y^n(s))\|^2 \, ds\right]$$

$$+ 2E\left[\int_{t_0}^{t_0+a} \|\sigma(s, y^n_r(s)) - \sigma(s, y^n(s))\|^2 \, ds\right]$$

$$\leq \varepsilon^3,$$

and by Chebyshev's inequality, we get

$$P(\|I^n_r(t) - I^n(t)\| > \varepsilon) < \varepsilon, \qquad r \geq r_0.$$

This, together with (4.2.12) and (4.2.14), shows that

$$P(\|y(t) - y_0 - I(t)\| > 6\varepsilon)$$
$$= P(\|(y(t) - y_r^r(t)) + (y_{r0} - y_0) - (I(t) - I^n(t))$$
$$+ (I_r^n(t) - I^n(t)) + (I_r(t) - I_r^n(t)) + (I_r^r(t) - I_r(t))\| > 6\varepsilon)$$
$$\leq P(\|y(t) - y_r^r(t)\| > \varepsilon) + P(\|y_{r0} - y_0\| > \varepsilon)$$
$$+ P(\|I(t) - I^n(t)\| > \varepsilon) + P(\|I_r^n(t) - I^n(t)\| > \varepsilon)$$
$$+ P(\|I_r(t) - I_r^n(t)\| > \varepsilon) + P(\|I_r^r(t) - I_r(t)\| > \varepsilon)$$
$$< 6\varepsilon \qquad \text{for} \quad n, r \geq \max(N(\varepsilon), r_0(\varepsilon)).$$

Since $\varepsilon > 0$ is arbitrary, this implies (4.2.13). Hence $y(t) \equiv x(t)$ is a solution of (4.2.1), and the proof is complete.

**Remark 4.2.1.**   We note from the proof of Theorem 4.2.1 that hypothesis $(H_1)$ can be replaced by a weaker hypothesis, namely,

$(H_1^*)$   There exist positive numbers $L$ and $M$ such that

$$\|f(t, x)\|^4 + \|\sigma(t, x)\|^4 \leq L + M\|x\|^4.$$

Also, one can prove the existence of a solution on $[t_0, \infty)$, provided that $f \in C[[t_0, \infty) \times R^n, R^n]$ and $\sigma \in C[[t_0, \infty) \times R^n, R^{nm}]$.

We shall next give sufficient conditions on $f$ and $\sigma$, in order to ensure the uniqueness of solutions of (4.2.1).

**Theorem 4.2.2.**   Assume that $f$ and $\sigma$ satisfy the relation

$$2(x - y)^T(f(t, x) - f(t, y))$$
$$+ \text{tr}\left[(\sigma(t, x) - \sigma(t, y))(\sigma(t, x) - \sigma(t, y))^T\right]$$
$$\leq g(t, \|x - y\|^2) \tag{4.2.15}$$

for $(t, x), (t, y) \in J \times R^n$, where $g \in C[J \times R^+, R]$ is concave, nondecreasing in $u$ for fixed $t$. Let $u(t) \equiv 0$ be the unique solution of

$$u' = g(t, u), \qquad u(t) = 0. \tag{4.2.16}$$

Then the stochastic system (4.2.1) has at most one solution on $J$.

*Proof.*   Let $x(t) = x(t, t_0, x_0)$ and $y(t) = y(t, t_0, x_0)$ be two solutions of (4.2.1) on $J$. Applying Itô's formula to $\|x(t) - y(t)\|^2$ and using the assumption of (4.2.15), we find

$$E[\|x(t) - y(t)\|^2]$$
$$= E\left[\int_{t_0}^t [2(x(s) - y(s))^T(f(s, x(s)) - f(s, y(s)))\right.$$
$$\left. + \text{tr}((\sigma(s, x(s)) - \sigma(s, y(s)))(\sigma(s, x(s)) - \sigma(s, y(s)))^T)] \, ds\right]$$
$$\leq E\left[\int_t^t g(s, \|x(s) - y(s)\|^2) \, ds\right].$$

Because of Jensen's inequality, the concavity of $g(t, u)$ now gives

$$E[\|x(t) - y(t)\|^2] \leq \int_{t_0}^t g(s, E[\|x(s) - y(s)\|^2]) \, ds, \qquad t \in J.$$

Set $m(t) = E[\|x(t) - y(t)\|^2]$ so that $m(t_0) = 0$. Define

$$u(t) = \int_{t_0}^t g(s, E[\|x(s) - y(s)\|^2]) \, ds.$$

Then we have

$$u' = g(t, m(t)).$$

Since $m(t) \leq u(t)$, the monotonicity of $g$ yields the differential inequality

$$u'(t) \leq g(t, u(t)), \qquad u(t_0) = 0.$$

By the scalar version of Theorem 1.7.1, we have

$$u(t) \leq r(t), \qquad t \in [t_0, t_0 + a],$$

where $r(t)$ is the maximal solution of (4.2.16) through $(t_0, 0)$. This implies that $m(t) \leq r(t)$, $t \in [t_0, t_0 + a]$, and consequently, the assumption $r(t) \equiv 0$ on $J$ proves the theorem.

**Remark 4.2.2.** We note that when the solution of (4.2.1) is unique, then the solution process is a Markov process and also a diffusion process.

The following result gives sufficient conditions under which (4.2.1) has a unique solution.

**Theorem 4.2.3.** Assume that

($H_1$)   the functions $f(t, x)$ and $\sigma(t, x)$ in (4.2.1) satisfy

$$\|f(t, x)\|^2 + \|\sigma(t, x)\|^2 \leq L^2(1 + \|x\|^2) \qquad \text{(growth condition)}$$

and

$$\|f(t, x) - f(t, y)\| + \|\sigma(t, x) - \sigma(t, y)\| \leq L\|x - y\|$$

$$\text{(Lipschitz condition)}$$

for all $(t, x), (t, y) \in J \times R^n$, where $L$ is some positive number.

($H_2$)   $x(t_0)$ is independent of $z(t)$ and satisfies $E[\|x_0\|^2] < \infty$.

Then there exists a unique solution of (4.2.1).

*Proof.* The proof of this result is given by the method of successive approximation. Let us define the zero approximation by $x_0(t) \equiv x_0$ for $t \in J$. Then let $x_n(t)$ be defined by

$$x_n(t) = x_0 + \int_{t_0}^t f(s, x_{n-1}(s)) \, ds + \int_{t_0}^t \sigma(s, x_{n-1}(s)) \, dz(s).$$

By virtue of the inequality $(x + y)^2 \leq 2(x^2 + y^2)$, Hölder's inequality, and $(H_1)$, we obtain

$$E[\|x_{n+1}(t) - x_n(t)\|^2] \leq 2\left[ E\left\| \int_{t_0}^t f(s, x_n(s)) - f(s, x_{n-1}(s))\, ds \right\|^2 \right]$$

$$+ 2E\left[ \left\| \int_{t_0}^t \sigma(s, x_n(s)) - \sigma(s, x_{n-1}(s))\, dz(s) \right\|^2 \right]$$

$$\leq (2aL^2 + 2L^2) \int_{t_0}^t E[\|x_n(s) - x_{n-1}(s)\|^2]\, ds$$

$$\leq M \int_{t_0}^t E[\|x_n(s) - x_{n-1}(s)\|^2]\, ds,$$

where $M = 2L^2(1 + a)$. Iterating this inequality, we find

$$E[\|x_{n+1}(t) - x_n(t)\|^2] \leq M^{n-1} \int_{t_0}^t \frac{(t - s)^{n-1}}{(n - 1)!} E[\|x_1(s) - x_0(s)\|^2]\, ds.$$

Since

$$E[\|x_1(t) - x_0(t)\|^2] \leq MaL^2(1 + E[\|x_0\|^2]),$$

it follows that there exists a constant $K$ such that

$$E[\|x_{n+1}(t) - x_n(t)\|^2] \leq K \frac{(Ma)^n}{n!}.$$

We note that

$$\left[ \sup_{t_0 \leq t \leq t_0 + a} \|x_{n+1}(t) - x_n(t)\| \right]$$

$$\leq \int_{t_0}^t \|f(s, x_n(s)) - f(s, x_{n-1}(s))\|\, ds$$

$$+ \sup_{t_0 \leq t \leq t_0 + a} \left[ \left\| \int_{t_0}^t (\sigma(s, x_n(s)) - \sigma(s, x_{n-1}(s)))\, dz(s) \right\| \right]$$

This, together with property (d) of the stochastic indefinite integral and $(H_1)$, yields

$$E\left[ \sup_{t_0 \leq t \leq t_0 + a} \|x_{n+1}(t) - x_n(t)\|^2 \right] \leq C_1 \frac{(aM)^{n-1}}{(n - 1)!},$$

where $C_1 = L^2(2T + 8)a$. From the convergence of

$$\sum_{n=1}^{\infty} P\left( \sup_{t_0 \leq t \leq t_0 + a} \|x_{n+1}(t) - x_n(t)\| > \frac{1}{n^2} \right) \leq \sum_{n=1}^{\infty} \frac{C_1(La)^{n-1}}{(n - 1)!} n^4$$

follows the uniform convergence of

$$x_0 + \sum_{n=1}^{\infty} [x_{n+1}(t) - x_n(t)] = x_n(t) \xrightarrow{\text{a.s.}} x(t)$$

on $J$. Since $x(t)$ is a limit of a sequence of nonanticipating functions and the uniform limit of continuous functions is continuous, it is itself nonanticipating and continuous. It remains to be shown that $x(t)$ satisfies Eq. (4.2.1). Because of the continuity of $f$ and $\sigma$, $(H_1)$, and the uniform convergence of $x_n$, we have

$$\left\| \int_{t_0}^t f(s, x_n(s)) - f(s, x(s)) \, ds \right\| \to 0 \quad \text{w.p. 1}$$

and

$$\int_{t_0}^t \|\sigma(s, x_n(s)) - \sigma(s, x(s))\|^2 \, ds \to 0 \quad \text{w.p. 1.}$$

From this and property (e) of the indefinite integral, it follows that $x(t)$ is a solution of (4.2.1).

We observe from the Lipschitz condition that

$$2(x - y)^T(f(t, x) - f(t, y))$$
$$+ \operatorname{tr}((\sigma(t, x) - \sigma(t, y))(\sigma(t, x) - \sigma(t, y))^T)$$
$$\leq 2\|x - y\| \|f(t, x) - f(t, y)\| + \|\sigma(t, x) - \sigma(t, y)\|^2$$
$$\leq 2L\|x - y\|^2 + L^2\|x - y\|^2 = L(2L + 1)\|x - y\|^2 \equiv g(t, \|x - y\|^2).$$

By the application of Theorem 4.2.2, the uniqueness of the solutions follows immediately and the proof is complete.

In many applications, the functions $f(t, x)$ and $\sigma(t, x)$ are not defined for all $x \in R^n$. It is therefore necessary to extend the existence theorem in order to cover such cases.

**Theorem 4.2.4.** Assume that all the hypotheses of Theorem 4.2.1 hold except that $C[J \times R^n, R^n]$ is replaced by $C[J \times B(x_0, \rho), R^n]$, where

$$B(x_0, \rho) = \{x \in R^n : |x_i - x_{i0}| \leq \rho_i, \rho_i > 0, i = 1, 2, \ldots, n\}$$

and $\rho = (\rho_1, \rho_2, \ldots, \rho_n)$. Then there exists a solution $x(t, t_0, x_0)$ of (4.2.1) that belongs to $B(x_0, \rho)$, with probability greater than or equal to $1 - M^*(t - t_0)/\|\rho\|^2$ on $[t_0, t_0 + a]$, where $M^* = 4M_1\sqrt{\alpha}$.

*Proof.* Define

$$\bar{f}(t, x) = f(t, \bar{x}) \quad \text{and} \quad \bar{\sigma}(t, x) = \sigma(t, \bar{x}),$$

where $\bar{x} = (\bar{x}_1, \bar{x}_2, \ldots, \bar{x}_n)$ and

$$\bar{x}_j = \begin{cases} \rho_j + x_{j_0} & \text{if} \quad x_j > \rho_j + x_{j_0}, \\ x_j & \text{if} \quad -\rho_j + x_{j_0} \le x_j \le \rho_j + x_{j_0}, \\ -\rho_j + x_{j_0} & \text{if} \quad x_j < -\rho_j + x_{j_0} \end{cases}$$

for $1 \le j \le n$. Note that $|\bar{x}_j - x_{j_0}| \le \rho_j$ for every $j = 1, 2, \ldots, n$. Furthermore, $\bar{f}(t, x)$ and $\bar{\sigma}(t, x)$ are continuous on $J \times R^n$ and satisfy $(H_1)$. Therefore, by Theorem 4.2.1, the modified problem

$$dx = \bar{f}(t, x) + \bar{\sigma}(t, x) \, dz(t) \tag{4.2.17}$$

has a solution $\bar{x}(t, t_0, x_0)$ which satisfies relation (4.2.4). Note that $\bar{x}(t) = x(t)$ as long as $\bar{x}(t) \in \bar{B}(x_0, \rho)$, where $x(t)$ is a solution of (4.2.1).

$$E[\|\bar{x}(t) - x_0\|^4] \le 4M_1(t - t_0) \qquad \text{for} \quad t \in [t_0, t_0 + a],$$

where $M_1$ is the number defined in the proof of Theorem 4.2.1. By Chebyshev's inequality,

$$P[\|x(t) - x_0\| > \|\rho\|] \le \frac{E[\|x(t) - x_0\|^2]}{\|\rho\|^2} \le \frac{4M_1(t - t_0)\sqrt{\alpha}}{\|\rho\|^2}.$$

Thus

$$P[\|x(t) - x_0\| < \|\rho\|] = P\{x(t) \in B(x_0, \rho)\} \ge 1 - \frac{M^*(t - t_0)}{\|\rho\|^2}$$

$$t \in [t_0, t_0 + a].$$

Hence the theorem is proved.

From the conclusion of Theorem 4.2.4, one can easily see that $x(t) \in B(x_0, \rho)$ for $t \in [t_0, t_0 + \varepsilon\|\rho\|^2/M^*]$ for arbitrary given $0 < \varepsilon < 1$.

**Problem 4.2.1.**   Prove the existence theorem (Theorem 4.2.1) by replacing hypothesis $(H_1)$ by $(H_1^{**})$, namely, for each $\rho$ there exist positive numbers $L(\rho)$ and $M(\rho)$ such that

$$\|f(t, x)\| + \|\sigma(t, x)\| \le L + M\|x\| \qquad \text{for} \quad (t, x) \in J \times \bar{B}(0, \rho).$$

(*Hint*:   Follow the proof of Theorem 4.2.4 and Theorem 3 in [20, pp. 45–47].)

### 4.3.   CONTINUOUS DEPENDENCE

In the formulation of a mathematical model for a physical or biological problem, we make errors in constructing the functions $f(t, x)$ and $\sigma(t, x)$ in (4.2.1) and errors in the initial conditions. For theoretical purposes it is sufficient to know that the change in the solution can be made arbitrary small by making the change in the differential equations and the initial values sufficiently small.

In the following theorem we establish the continuous dependence of solutions on the parameter.

**Theorem 4.3.1.**  Assume that

$(H_1)_\lambda$  the $m$-column vectors of $\sigma(t, x, \lambda)$ and $f(t, x, \lambda)$ belong to $C[J \times R^n \times \Lambda, R^n]$, $J = [t_0, t_0 + a]$, and $\Lambda$ is a parameter set;

$(H_2)_\lambda$  there exist positive numbers $L$ and $M$ such that

$$\|f(t, x, \lambda)\| + \|\sigma(t, x, \lambda)\| \le L + M\|x\|$$

for $(t, x, \lambda) \in J \times R^n \times \Lambda$;

$(H_3)_\lambda$  there exists a function $g \in C[J \times R_+, R]$, which is concave and nondecreasing in $u$ for fixed $t \in J$, such that

$$2[(x - y)^T(f(t, x, \lambda) - f(t, y, \lambda)) + \|\sigma(t, x, \lambda) - \sigma(t, y, \lambda)\|^2]$$
$$\le g(t, \|x - y\|^2) \tag{4.3.1}$$

for all $(t, x, \lambda), (t, y, \lambda) \in J \times R^n \times \Lambda$;

$(H_4)_\lambda$  $x(t_0, \lambda)$ is independent of $z(t)$ and

$$\lim_{\lambda \to \lambda_0} E[\|x(t_0, \lambda) - x(t_0, \lambda_0)\|^2] = 0, \tag{4.3.2}$$

$(H_5)_\lambda$  $E[\|x(t_0, \lambda)\|^4] \le c_1$ for some constant $c_1 > 0$;

$(H_6)_\lambda$  for any $\varepsilon > 0$, $N > 0$,

$$\lim_{\lambda \to \lambda_0} P\left[\sup_{\|x\| \le N} (\|f(t, x, \lambda) - f(t, x, \lambda_0)\|\right.$$

$$\left. + \|\sigma(t, x, \lambda) - \sigma(t, x, \lambda_0)\|) > \varepsilon\right] = 0; \tag{4.3.3}$$

$(H_7)$  $u(t) \equiv 0$ is the unique solution of

$$u' = g(t, u) \tag{4.3.4}$$

with $u(t_0) = 0$ and the solutions $u(t, t_0, u_0)$ of (4.3.4) with $u(t_0) = u_0$ are continuous with respect to $(t_0, u_0)$.

Then given $\varepsilon > 0$, there exists a $\delta(\varepsilon) > 0$ such that for every $\lambda$, $\|\lambda - \lambda_0\| < \delta(\varepsilon)$, the differential system

$$dx = f(t, x, \lambda)\, dt + \sigma(t, x, \lambda)\, dz(t), \qquad x(t_0, \lambda) = x_0(\lambda) \tag{4.3.5}$$

admits a unique solution $x(t, \lambda) = x(t, t_0, x(t_0, \lambda))$ satisfying

$$(E[\|x(t, \lambda) - x(t, \lambda_0)\|^2])^{1/2} < \varepsilon \qquad \text{for all} \quad t \in J. \tag{4.3.6}$$

*Proof.*   Under hypotheses $(H_1)_\lambda, (H_2)_\lambda, (H_4)_\lambda$, and $(H_5)_\lambda$, the initial value problem (4.3.5) has at least one solution $x(t, \lambda) = x(t, t_0, x(t_0, \lambda))$ by Theorem 4.2.1. That this solution is unique follows by Theorem 4.2.2 in view of hypotheses $(H_3)_\lambda$ and $(H_7)_\lambda$. Moreover, $x(t, \lambda)$ satisfies the estimate (4.2.4) for all $\lambda \in \Lambda$. By applying Itô's formula to the function $\|x - y\|^2$, assumption $(H_3)_\lambda$, and concavity of $g$ in $u$, we get

$$E[\|x(t, \lambda) - x(t, \lambda_0)\|^2]$$

$$= E[\|x(t_0, \lambda) - x(t_0, \lambda_0)\|^2] + E\left[\int_{t_0}^t [2(x(s, \lambda) - x(s, \lambda_0))^{\mathrm{T}}\right.$$

$$\times (f(s, x(s, \lambda), \lambda) - f(s, x(s, \lambda_0), \lambda_0))$$

$$+ \mathrm{tr}((\sigma(s, x(s, \lambda), \lambda) - \sigma(s, x(s, \lambda_0), \lambda_0))$$

$$\left. \times (\sigma(s, x(s, \lambda), \lambda) - \sigma(s, x(s, \lambda_0), \lambda_0))^{\mathrm{T}})] \, ds\right]$$

$$\le E[\|x(t_0, \lambda) - x(t_0, \lambda_0)\|^2]$$

$$+ \int_{t_0}^t g(s, E[\|x(s, \lambda) - x(s, \lambda_0)\|^2]) \, ds$$

$$+ E\left[\int_{t_0}^t 2[(x(s, \lambda) - x(s, \lambda_0))^{\mathrm{T}}\right.$$

$$\times (f(s, x(s, \lambda_0), \lambda) - f(s, x(s, \lambda_0), \lambda_0))$$

$$\left. + \|\sigma(s, x(s, \lambda_0), \lambda) - \sigma(s, x(s, \lambda_0), \lambda_0)\|^2] \, ds\right].$$

Using hypotheses $(H_2)_\lambda, (H_4)_\lambda, (H_6)_\lambda$ and the Lebesgue dominated convergence theorem, relations (4.2.4) and (4.3.2) give the inequality

$$E[\|x(t, \lambda) - x(t, \lambda_0)\|^2] \le \eta + \int_{t_0}^t g(s, E[\|x(s, \lambda) - x(s, \lambda_0)\|^2) \, ds, \quad (4.3.7)$$

for $t \in J$, whenever $\|\lambda - \lambda_0\| \le \delta(\eta)$, $\eta > 0$ being arbitrary. Set $m(t) = E[\|x(t, \lambda) - x(t, \lambda_0)\|^2]$ and

$$u(t) = \eta + \int_{t_0}^t g(s, m(s)) \, ds, \qquad u(t_0) = \eta.$$

Then relation (4.3.7), together with the nondecreasing property of $g$, shows that

$$u'(t) \le g(t, u(t)), \qquad u(t_0) = \eta.$$

Hence by the scalar version of Theorem 1.7.1, we have

$$u(t) \le r(t, t_0, u_0), \qquad t \in J,$$

where $r(t, t_0, \eta)$ is the maximal solution of (4.3.4) through $(t_0, u_0)$ with $u_0 = \eta$. This, together with (4.3.7) and the definition of $u(t)$, yields

$$m(t) \leq r(t, t_0, \eta), \qquad t \in J.$$

Since the solutions $u(t, t_0, u_0)$ of (4.3.4) are assumed to be continuous with respect to $u_0$, it follows that

$$\lim_{\eta \to 0} r(t, t_0, \eta) = r(t, t_0, 0),$$

and by hypothesis, $r(t, t_0, 0) \equiv 0$. This, in view of the definition of $m(t)$, yields

$$\lim_{\lambda \to \lambda_0} E[\|x(t, \lambda) - x(t, \lambda_0)\|^2] = 0,$$

which establishes the theorem.

Now we shall state a theorem which ensures the continuous dependence of solutions of (4.2.1) on the initial conditions $(t_0, x_0)$.

**Theorem 4.3.2.** Assume that the hypotheses of Theorems 4.2.1 and 4.2.2 hold. Furthermore, suppose that the solutions $u(t, t_0, u_0)$ of (4.3.4) through every point $(t_0, u_0)$ are continuous with respect to $(t_0, u_0)$. Then the solutions $x(t, t_0, x_0)$ of (4.2.1) are unique and continuous with respect to the initial conditions $(t_0, x_0)$.

*Proof.* The existence and uniqueness of the solutions $x(t, x_0, x_0)$ of (4.2.1) follow from Theorems 4.2.1 and 4.2.2. The proof of continuous dependence can be formulated based on the proof of Theorem 4.3.1. We shall merely indicate the necessary steps in the proof.

Let $x(t, t_0, x_0)$ and $x(t, t_1, y_0)$ be solutions of (4.2.1) through $(t_0, x_0)$ and $(t_1, y_0)$, respectively, By Itô's formula we get, assuming $t_1 \geq t_0$,

$$E[\|x(t, t_0, x_0) - x(t, t_1, y_0)\|^2]$$

$$= E[\|x(t_1, t_0, x_0) - y_0\|^2] + E\left[ \int_{t_1}^t [2(x(s, t_0, x_0) - x(s, t_1, y_0))^\mathsf{T} \right.$$

$$\times (f(s, x(s, t_0, x_0)) - f(s, x(s, t_1, y_0)))]$$
$$+ \text{tr}[(\sigma(s, x(s, t_0, x_0)) - \sigma(s, x(s, t_1, y_0)))$$

$$\left. \times (\sigma(s, x(s, t_0, x_0)) - \sigma(s, x(s, t_1, y_0)))^\mathsf{T}] \, ds \right].$$

This, together with (4.2.15) and the fact that

$$E[\|x(t_1, t_0, x_0) - y_0\|^2] \leq 2E[\|x(t_1, t_0, x_0) - x(t_0, t_0, x_0)\|^2] + E[\|x_0 - y_0\|^2],$$

yields

$$E[\|x(t,t_0,x_0) - x(t,t_1,y_0)\|^2]$$
$$\leq 2E[\|x(t_1,t_0,x_0) - x(t_0,t_0,x_0)\|^2] + E[\|x_0 - y_0\|^2]$$
$$+ E\left[\int_{t_1}^{t} g(s, \|x(s,t_0,x_0) - x(s,t_1,y_0)\|^2)\,ds\right].$$

Now the rest of the proof can be constructed by following the proof of the Theorem 4.3.1. We leave the details to the reader.

**Remark 4.3.1.**   If hypothesis $(H_5)_\lambda$ in Theorem 4.3.1 is replaced by

$(H_5^*)_\lambda$   $E[\|x(t,\lambda)\|^2] \leq c_1$ for some constant $c_1 > 0$,

then the conclusion of the theorem remains valid.

We now turn to the question of the differentiability of the solution of (4.2.1) with respect to the initial conditions $(t_0,x_0)$. We shall consider the mean square derivatives $(\partial x/\partial x_0)(t,t_0,x_0)$ and $(\partial x/\partial t_0)(t,t_0,x_0)$.

**Theorem 4.3.3.**   Assume that

$(H_1)$   the functions $f(t,x)$ and $\sigma(t,x)$ in (4.2.1) satisfy the relations

$$\|f(t,x)\|^2 + \|\sigma(t,x)\|^2 \leq L^2(1 + \|x\|^2),$$
$$\|f(t,x) - f(t,y)\| + \|\sigma(t,x) - \sigma(t,y)\| \leq L\|x - y\|$$

for $(t,x),(t,y) \in J \times R^n$, where $L$ is a positive constant;

$(H_2)$   $x(t_0)$ is independent of $z(t)$ and $E[\|x(t_0)\|^2] < \infty$;

$(H_3)$   $f(t,x)$ and $\sigma(t,x)$ are continuously differentiable with respect to $x$ for fixed $t \in J$ and their derivatives $f_x(t,x)$ and $\sigma_x(t,x)$ are continuous in $(t,x) \in J \times R^n$ and satisfy the condition

$$\|f_x(t,x)\| + \|\sigma_x(t,x)\| \leq C$$

for some positive constant $C$.

Then

(a)   The mean square derivative $(\partial/\partial x_{k_0})x(t,t_0,x_0)$ exists for $k = 1, 2, \ldots, n$ and satisfies the Itô-type system of linear differential equations

$$dy = f_x(t,x(t))y\,dt + \sigma_x(t,x(t))y\,dz \qquad (4.3.8)$$

where $y(t_0) = e_k = (e_k^1, e_k^2, \ldots, e_k^k, \ldots, e_k^n)$ is an $n$-vector such that $e_k^j = 0$, $j \neq k$, and $e_k^k = 1$, and $x_{k_0}$ is the $k$th component of $x_0$; moreover,

$$d\phi = f_x(t,x(t))\phi\,dt + \sigma_x(t,x(t))\phi\,dz \qquad (4.3.9)$$

where $\phi = \phi(t,t_0,x_0) = (\partial x/\partial x_0)(t,t_0,x_0)$;

(b)   the mean square derivative $(\partial/\partial t_0) x(t, t_0, x_0)$ exists and satisfies (4.3.8) with $(\partial/\partial t_0) x(t, t_0, x_0) = -\Phi(t, t_0, x_0) f(t_0, x_0)$.

*Proof.*   From hypotheses $(H_1)$ and $(H_2)$ and by the method of successive approximation, it has been shown that (4.2.1) has a unique solution $x(t) = x(t, t_0, x_0)$. Moreover, by Theorem 4.3.2, solutions are continuously dependent with respect to $(t_0, x_0)$. For small $\lambda$, let $x(t, \lambda) = x(t, t_0, x_0 + \lambda e_k)$ be a solution process through $(t_0, x_0 + \lambda e_k)$. From the continuous dependence on initial conditions, it is clear that

$$x(t, \lambda) \xrightarrow{\text{m.s.}} x(t) \qquad \text{as} \quad \lambda \to 0 \quad \text{uniformly on} \quad J. \qquad (4.3.10)$$

Set

$$\Delta x(t, \lambda) = [x(t, \lambda) - x(t)]/\lambda, \qquad \Delta x(t_0, \lambda) - e_k, \qquad \lambda \neq 0,$$
$$x = x(t), \qquad \text{and} \qquad y = x(t, \lambda).$$

This, together with the application of Lemma 2.6.1, yields

$$f(t, y) - f(t, x) = \int_0^1 f_x(t, sy + (1 - s)x)(y - x)\, ds \qquad (4.3.11)$$

and

$$\sigma(t, y) - \sigma(t, x) = \int_0^1 \sigma_x(t, sy + (1 - s)x)(y - x)\, ds. \qquad (4.3.12)$$

Let us denote

$$F(t, x(t), \lambda) = \int_0^1 f_x(t, sy + (1 - s)x)\, ds$$

and

$$G(t, x(t), \lambda) = \int_0^1 \sigma_x(t, sy + (1 - s)x)\, ds.$$

This, together with $(H_3)$, yields

$$\|F(t, x(t), \lambda)\| \leq C \qquad \text{and} \qquad \|G(t, x(t), \lambda)\| \leq C. \qquad (4.3.13)$$

Furthermore, from (4.3.10), we note that $F(t, x, \lambda)$ and $G(t, x, \lambda)$ are continuous in $(t, x, \lambda)$ and satisfy

$$F(t, x, \lambda) \xrightarrow{\text{a.s.}} f_x(t, x) \qquad \text{and} \qquad G(t, x, \lambda) \xrightarrow{\text{a.s.}} \sigma_x(t, x) \qquad (4.3.14)$$

as $\lambda \to 0$ uniformly on $J$. On the other hand,

$$d(y - x) = [f(t, y) - f(t, x)]\, dt + [\sigma(t, y) - \sigma(t, x)]\, dz.$$

This, together with (4.3.11) and (4.3.12) and the definitions of $\Delta x(t, \lambda)$, $F(t, x, \lambda)$, and $G(t, x, \lambda)$, yields

$$d\,\Delta x(t, \lambda) = F(t, x, \lambda)\,\Delta x(t, \lambda)\, dt + G(t, x, \lambda)\,\Delta x(t, \lambda)\, dz,$$
$$\Delta x(t_0, \lambda) = e_k. \qquad (4.3.15)$$

From (4.3.13), (4.3.14), (4.3.15), Chebyshev's inequality, the continuous dependence of the solution on the initial state, and the definition of $\Delta x(t, \lambda)$, it follows that

$$\|F(t, x, \lambda)y\| + \|G(t, x, \lambda)y\| \leq L + M\|y\|,$$

$$\|F(t, x, \lambda)y_1 - F(t, x, \lambda)y_2\| + \|G(t, x, \lambda)y_1 - G(t, x, \lambda)y_2\|$$

$$\leq C\|y_1 - y_2\|, \quad \lim_{\lambda \to \lambda_0} P\left( \sup_{\|y\| \leq N} (\|F(t, x(t, \lambda), \lambda) - F(t, x(t, \lambda_0), \lambda_0)\| \right.$$

$$\left. + \|G(t, x(t, \lambda), \lambda) - G(t, x(t, \lambda_0), \lambda_0)\|\right) > \varepsilon \right) = 0,$$

and

$$\lim_{\lambda \to 0} E[\|x(t_0, \lambda) - x(t_0, \lambda_0)\|] = \lim_{\lambda \to \lambda_0} |\lambda - \lambda_0| = 0,$$

where $M = 2C$, $\varepsilon > 0$ is any number, $\lambda_0 \in \Lambda$, and $L$ is any positive constant. By the application of Theorem 4.2.3, (4.3.15) has a unique solution $\Delta x(t, \lambda)$ with $\Delta x(t_0, \lambda) = e_k$ on $J$. As noted in the proof of Theorem 4.2.3, we have

$$2(y_1 - y_2)^\mathsf{T} F(t, x, \lambda)y_1 - F(t, x, \lambda)y_2 + \|G(t, x, \lambda)y_1 - G(t, x, \lambda)y_2\|^2$$
$$\leq C(C + 2)\|y_1 - y_2\|^2 \equiv g(t, \|y_1 - y_2\|^2).$$

Therefore, by the application of Theorem 4.3.1 in the context of Remark 4.3.1, $\lim_{\lambda \to 0} \Delta x(t, \lambda)$ exists in the mean square, and it is a solution of (4.3.8). From the definition of $\Delta x(t, \lambda)$, the limit of $\Delta x(t, \lambda)$ as $\lambda \to 0$ is equal to the partial derivative of $x(t, t_0, x_0)$ with respect to the $k$th component of $x_0$. This is true for all $k = 1, 2, \ldots, n$. Thus $(\partial/\partial x_0)x(t, t_0, x_0)$ is the fundamental solution process of (4.3.8) and satisfies the Itô-type random matrix differential equation

$$dY = f_x(t, x(t))Y \, dt + \sigma_x(t, x(t))Y \, dz, \qquad \Phi(t_0) = I, \qquad (4.3.16)$$

where $I$ is a random identity matrix and $x(t) = x(t, t_0, x_0)$ is the solution process of (4.2.1). We note that the fundamental solution of (4.3.8) is denoted by $(\partial/\partial x_0)x(t, t_0, x_0) = \Phi(t, t_0, x_0)$. This establishes the proof of (a).

To prove part (b), without loss of generality, take $t_0 \leq s < s + \lambda \leq t \leq t_0 + b$, and define

$$\Delta x(t, \lambda) = \frac{x(t, \lambda) - x(t)}{\lambda}, \qquad \Delta x(s, \lambda) = \frac{x(s, \lambda) - x_0}{\lambda},$$

where $x(t, \lambda) = x(t, s + \lambda, x_0)$ and $x(t) = x(t, s, x_0)$ are solution processes through $(s + \lambda, x_0)$ and $(s, x_0)$, respectively. Again, by imitating the proof of the part (a), one can conclude that the derivative $(\partial/\partial t_0)x(t, t_0, x_0)$ exists in the mean square sense, and it is the solution of (4.3.8) whenever

$$\lim_{\lambda \to 0} \Delta x(s, \lambda) = -\lim_{\lambda \to 0} \frac{1}{\lambda} \left[ \int_s^{s+\lambda} f(u, x(u)) \, du + \sigma(s, x(u)) \, dz(u) \right]$$

exists in the mean square sense. The proof is complete.

The following example shows that under certain conditions the solution process $x(t, t_0, x_0)$ is twice differentiable with respect to $x_0$.

**Example 4.3.1.** In addition to the hypotheses of Theorem 4.3.3, we assume that $f(t, x)$ and $\sigma(t, x)$ are twice continuously differentiable and their derivatives $f_{xx}(t, x)$ and $\sigma_{xx}(t, x)$ satisfy the inequality

$$\| f_{xx}(t, x) \| + \| \sigma_{xx}(t, x) \| \le L_2(1 + \|x\|^{m_2}) \tag{4.3.17}$$

for some positive numbers $m_2$ and $L_2$. Then the mean square derivative $(\partial^2 / \partial x_0 \, \partial x_0) x(t, t_0, x_0)$ exists and satisfies the Itô-type stochastic matrix differential equation

$$dY = \left[ f_{xx}(t, x(t)) \Phi^{\mathrm{T}}(t, t_0, x_0) \Phi(t, t_0, x_0) + f_x(t, x(t)) Y \right] dt$$
$$+ \left[ \sigma_{xx}(t, x(t)) \Phi^{\mathrm{T}}(t, t_0, x_0) \Phi(t, t_0, x_0) + \sigma_x(t, x(t)) Y \right] dz$$

where $\Phi(t, t_0, x_0)$ is the fundamental matrix solution of (4.3.8).

*Proof.* The validity of the above statement can be established by following the argument used in the proof of Theorem 4.3.3. However, we note that relation (4.3.17) is needed first to show that $\Delta x(t, \lambda)$ is bounded which in turn implies that

$$\left[ \frac{1}{\lambda} [f(t, x(t, \lambda)) - f(t, x(t))] - f_x(t, x(t)) \right]$$

and

$$\left[ \frac{1}{\lambda} [\sigma(t, x(t, \Delta)) - \sigma(t, x(t))] - \sigma_x(t, x(t)) \right]$$

converge to zero in the mean square uniformly on $J$. Further details are left as an exercise.

## 4.4. THE METHOD OF VARIATION OF PARAMETERS

Consider a system of deterministic differential equations

$$x' = f(t, x), \qquad x(t_0) = x_0, \tag{4.4.1}$$

where $f \in C[R_+ \times R^n, R^n]$. The stochastic differential system of Itô-type (4.2.1), namely,

$$dy = f(t, y) \, dt + \sigma(t, y) \, dz(t), \qquad y(t_0) = x_0, \tag{4.4.2}$$

can be viewed as a perturbed system relative to (4.4.1) with a constantly acting stochastic perturbation. With this understanding, we shall obtain a nonlinear variation of constants formula for the solutions of (4.2.1).

**Theorem 4.4.1.**   Assume that $f(t, x)$ is twice continuously differentiable with respect to $x$ for each $t \in R_+$ and that $y(t) = y(t, t_0, x_0)$ and $x(t) = x(t, t_0, x_0)$ are solutions of (4.2.1) and (4.4.1), respectively, existing for $t \geq t_0$. Then

$$y(t, t_0, x_0) = x(t, t_0, x_0) + \int_{t_0}^t \frac{\partial x}{\partial x_0}(t, s, y(s))\sigma(s, y(s))\, dz(s)$$

$$+ \frac{1}{2} \int_{t_0}^t tr\left(\frac{\partial x}{\partial x_0 \partial x_0}(t, s, y(s))b(s, y(s))\right)\, ds, \qquad t \geq t_0, (4.4.3)$$

where $b(t, y) = \sigma(t, y)\sigma^T(t, y)$.

*Proof.*   Consider $x(t, s, y(s))$. Under the assumption on $f$, it is known (see Hartman [24]) that the solution $x(t, t_0, x_0)$ of (4.4.1) is continuously differentiable with respect to $t_0$ and twice continuously differentiable with respect to $x_0$. As a result, applying Itô's formula to $x(t, s, y(s))$, we have

$$dx(t, s, y(s)) = \frac{\partial x}{\partial t_0}(t, s, y(s))\, ds + \frac{\partial x}{\partial x_0}(t, s, y(s))\, dy(s)$$

$$+ \frac{1}{2}\left(tr\left(\frac{\partial^2 x}{\partial x_0 \partial x_0}(t, s, y(s))\sigma(s, y(s))\sigma^T(s, y(s))\right)\right)\, ds. \quad (4.4.4)$$

We also know that

$$\frac{\partial x}{\partial t_0}(t, t_0, x_0) = -\frac{\partial x}{\partial x_0}(t, t_0, x_0)f(t_0, x_0),$$

and consequently, we derive from (4.4.4)

$$dx(t, s, y(s)) = \frac{\partial x}{\partial x_0}(t, s, y(s))\sigma(s, y(s))\, dz(s)$$

$$+ \frac{1}{2}\left(tr\left(\frac{\partial^2 x}{\partial x_0 \partial x_0}(t, s, y(s))b(s, y(s))\right)\right)\, ds.$$

Integrating this from $t_0$ to $t$, the desired result (4.4.3) follows.

**Corollary 4.4.1.**   If $f(t, x) = Ax$, where $A$ is an $n \times n$ matrix, that is, $f$ is linear, formula (4.4.3) reduces to the form

$$y(t, t_0, x_0) = x(t, t_0, x_0) + \int_{t_0}^t e^{A(t-s)}\sigma(s, y(s))\, dz(s), \qquad t \geq t_0.$$

**Problem 4.4.1.** By applying Itô's formula to $\|x(t, s, y(s))\|^p$, $p \geq 1$, find a variation of constants formula for $\|y(t, t_0, x_0)\|^p$ similar to relation (4.4.3). Deduce from it the corresponding formula when $f(t, x)$ is linear.

Consider now the Itô system

$$dx = \sigma(t, x)\, dz(t), \qquad x(t_0) = x_0, \tag{4.4.5}$$

to be an unperturbed system so that (4.4.2) can be viewed as a perturbed system with the deterministic perturbation $f(t, x)\, dt$. We can also derive a nonlinear variation of constants formula relative to this setup.

**Theorem 4.4.2.** Suppose that $\sigma(t, x)$ is twice continuously differentiable with respect to $x$ for each $t \in R_+$ and that $x(t) = x(t, t_0, x_0)$ and $y(t) = y(t, t_0, x_0)$ are solutions of (4.4.5) and (4.4.2), respectively, existing for $t \geq t_0$. Then

$$y(t, t_0, x_0) = x\left[ t, t_0, x_0 + \int_{t_0}^t \psi^{-1}(s, t_0, u(s)) f(s, y(s, t_0, x_0))\, ds \right], \qquad t \geq t_0,$$

where $(\partial x/\partial x_0)(t, t_0, x_0) \equiv \psi(t, t_0, x_0)$ and $u(t)$ is a solution of

$$u'(t) = \psi^{-1}(t, t_0, u(t)) f(t, x(t, t_0, u(t))), \qquad u(t_0) = x_0.$$

*Proof.* Let $x(t, t_0, x_0)$ be the solution of (4.4.5) for $t \geq t_0$. The method of variation of constants requires that we determine a process $u(t)$ such that

$$y(t, t_0, x_0) = x(t, t_0, u(t)), \qquad u(t_0) = x_0,$$

is a solution of (4.4.2). Since we know that $x(t, t_0, x_0)$ is twice continuously differentiable with respect to $x_0$ and that $\psi^{-1}(t, t_0, x_0)$ exists, where $\psi(t, t_0, x_0) = (\partial x/\partial x_0)(t, t_0, x_0)$ is the solution of the variational system relative to (4.4.5), we have by Itô's formula

$$dy = dx + \psi\, du + d\psi\, du + \frac{1}{2} \mathrm{tr}\left( \frac{\partial^2 x}{\partial x_0\, \partial x_0}(t, t_0, u(t))(du)^{\mathrm{T}}(du) \right)$$

Let us suppose that $u(t)$ satisfies

$$du = g(t, u)\, dt + G(t, u)\, dz(t), \qquad u(t_0) = x_0,$$

and find $g$ and $G$. Thus we have

$$f(t, y)\, dt = \psi\, du + \sigma_x(t, x)\psi G\, dt$$

$$+ \frac{1}{2} \mathrm{tr}\left( \frac{\partial^2 x}{\partial x_0\, \partial x_0}(t, t_0, u)G(t, u)^{\mathrm{T}}G(t, u) \right) dt.$$

Consequently, we see that $G \equiv 0$ and that

$$g(t, u) = \psi^{-1}(t, t_0, u(t)) f(t, y).$$

Hence it follows that

$$y(t, t_0, x_0) = x\left[t, t_0, x_0 + \int_{t_0}^t \psi^{-1}(s, t_0, u(s)) f(s, y(s, t_0, x_0)) \, ds\right],$$

for $t \geq t_0$.

As an illustration of Theorem 4.4.2, consider $\sigma(t, x) = \lambda(t)x$ in (4.4.5). Then we know by Example 4.1.3 that

$$x(t, t_0, x_0) = x_0 \exp\left[-\tfrac{1}{2}\int_{t_0}^t \lambda^2(s) \, ds + \int_{t_0}^t \lambda(s) \, dz(s)\right].$$

Hence

$$\psi(t, t_0) = \exp\left[-\tfrac{1}{2}\int_{t_0}^t \lambda^2(s) \, ds + \int_{t_0}^t \lambda(s) \, dz(s)\right],$$

and consequently any solution $y(t)$ of (4.4.2) is of the form

$$y(t) = \psi(t, t_0)$$
$$\times \left[x_0 + \int_{t_0}^t \exp\left(\tfrac{1}{2}\int_{t_0}^s \lambda^2(\xi) \, d\xi - \int_{t_0}^s \lambda(\xi) \, dz(\xi)\right) f(s, y(s)) \, ds\right]$$

**Problem 4.4.2.**   Consider the perturbed system

$$dy = [\sigma(t, y) + F(t, y)] \, dz + f(t, y) \, dt$$

relative to (4.4.5). Obtain the variation of constants formula for the solution $y(t)$, assuming that $\sigma(t, x)$ is twice continuously differentiable in $x$.

## 4.5.  STOCHASTIC DIFFERENTIAL INEQUALITIES

In this section we shall present some basic results on stochastic differential inequalities of Itô-type.

The following lemmas play an important role in our discussion.

**Lemma 4.5.1.**   Assume that

(i)   $m(t)$ is a solution of

$$dm = g(t, m) \, dt + G(t, m) \, dz(t), \qquad m(t_0) = m_0,$$

where $g, G \in C[R_+ \times R, R]$ and $z(t)$ is a normalized Wiener process;

(ii)   $|G(t, u)|^2 \leq h(|u|)$ for $(t, u) \in R_+ \times R$, where $h \in C[R_+, R_+]$ such that $h(0) = 0$, $h(u)$ is increasing in $u$, and

$$\int_{0^+} \frac{ds}{h(s)} = \infty.$$

Then we have

$$E(|m(t)|) \leq E(|m_0|) + E\left(\int_{t_0}^t |g(s, m(s))| \, ds\right), \qquad t \geq t_0.$$

*Proof.* Let $a_0 = 1 > a_1 > a_2 > \cdots > a_n \to 0$ as $n \to \infty$. Define for $n = 1, 2, \ldots, \int_{a_n}^{a_{n-1}} ds/h(s) = n$. Then there exists a twice continuously differentiable function $H_n(u)$ on $R_+$ such that $H_n(0) = 0$,

$$H_n'(u) = \begin{cases} 0, & 0 \leq u \leq a_n, \\ \text{between 0 and 1,} & a_n < u < a_{n-1}, \\ 1, & u \geq a_{n-1}, \end{cases}$$

and

$$H_n''(u) = \begin{cases} 0, & 0 \leq u \leq a_n, \\ \text{between 0 and } \dfrac{2}{nh(u)}, & a_n < u < a_{n-1}, \\ 0, & u \geq a_{n-1}. \end{cases}$$

We then extend $H_n(u)$ to $(-\infty, \infty)$ symmetrically, that is, $H_n(u) = H_n(|u|)$. Clearly, $H_n(u)$ is a twice continuously differentiable function on $(-\infty, \infty)$, and $H_n(u) \to |u|$ as $n \to \infty$.

Using Itô's formula, we get

$$H_n(m(t)) = H_n(m_0) + \int_{t_0}^t H_n'(m(s))G(s, m(s)) \, dz(s)$$
$$+ \int_{t_0}^t H_n'(m(s))g(s, m(s)) \, ds$$
$$+ \tfrac{1}{2} \int_{t_0}^t H_n''(m(s)) [G(s, m(s))]^2 \, ds$$
$$= H_n(m_0) + I_1 + I_2 + I_3.$$

It is obvious that $E(I_1) = 0$. Since $|H_n'(u)| \leq 1$, we have

$$E(I_2) \leq E\left(\int_{t_0}^t |g(s, m(s))| \, ds\right), \qquad t \geq t_0.$$

Also,

$$E(|I_3|) \leq \tfrac{1}{2}E\left(\int_{t_0}^t H_n''(|m(s)|)h(|m(s)|) \, ds\right)$$
$$\leq \tfrac{1}{2}(t - t_0) \max_{a_n \leq u \leq a_{n-1}} H_n''(u)h(|u|)$$
$$\leq \tfrac{1}{2}(t - t_0)\frac{2}{n} \to 0 \qquad \text{as} \quad n \to \infty.$$

Consequently, we obtain

$$E(|m(t)|) \leq E(|m_0|) + E \int_{t_0}^{t} |g(s, m(s)| \, ds, \qquad t \geq t_0,$$

and the proof is complete.

An extension of Lemma 4.5.1 to Itô-type systems of differential equations will be needed to derive component-wise estimates. We merely state such a result since its proof is similar, with appropriate modifications, to the proof of Lemma 4.5.1. As usual, inequalities between two vectors are to be understood component-wise. If $m \in R^n$, we use the notation $|m| = (|m_1|, |m_2|, \ldots, |m_n|)$.

**Lemma 4.5.2.**   Assume that

(i)   $m(t)$ is a solution of the Itô-type system

$$dm = g(t, m) \, dt + G(t, m) \, dz(t), \qquad m(t_0) = m_0,$$

where $g \in C[R_+ \times R^n, R^n]$, $G \in C[R_+ \times R^n, R^{nm}]$, and $z(t)$ is a normalized $m$-vector Wiener process;

(ii)   $\sum_{j=1}^{m} |G_{ij}(t, u)|^2 \leq h_i(|u_i|)$ for $(t, u) \in R_+ \times R^n$, where for each $i = 1, 2, \ldots, n, h_i \in C[R_+, R_+], h_i(0) = 0, h_i(s)$ is increasing in $s$, and $\int_{0^+} ds/h_i(s) = \infty$.

Then we have

$$E[|m(t)|] \leq E[|m_0|] + E\left( \int_{t_0}^{t} |g(s, m(s))| \, ds \right), \qquad t \geq t_0.$$

Let us introduce the notion of quasi-positivity of a function.

**Definition 4.5.1.**   A function $g \in C[R_+ \times R^n, R^n]$ is said to be strictly quasi-positive in $u \in R^n$ for each $t \in R_+$ if $u \geq 0$ and $u_i = 0$ imply $g_i(t, u) > 0$ for each $t \in R_+$.

We are now in position to prove the following result concerning the nonnegativity of solutions of Itô systems, which result is a useful tool in proving theorems on differential inequalities.

**Theorem 4.5.1.**   Let the assumptions of Lemma 4.5.2 hold. Suppose further that $g(t, u)$ is strictly quasi-positive in $u$ for each $t > t_0$. Then $m(t_0) = m_0 \geq 0$ implies $m(t) \geq 0$ for $t \geq t_0$ a.s.

*Proof.*   Let $\partial R^n_+$ denote the boundary of $R^n_+$. For any $m_0 \in \partial R^n_+$, one can find a nonempty index subset $I_0$ of $I = \{1, 2, \ldots, n\}$, depending on $m_0$, such that $m_{0i} = 0$ for $i \in I_0$ and $m_{0i} > 0$ for $i \in I \backslash I_0$, w.p. 1. We first prove that

$$P(\text{there exists a } t > t_0 : m(s) \geq 0 \text{ for } s \in [t_0, t]) = 1. \qquad (4.5.1)$$

To prove this, set

$$\tau_0 = \inf \bigcup_{i \in I_0} [s : g_i(s, m(s)) \le 0]$$

and

$$\tau_1 = \inf \bigcup_{i \in I - I_0} [s : m_i(s) < 0].$$

Because of the quasi-positivity of $g$ in $u$, $g_i(t, m_0) > 0$ for $i \in I_0$, $t > t_0$. This, together with the continuity of $g$ and the sample continuity of $m(t)$, gives $P(\tau > t_0) = 1$, where $\tau = \min(\tau_0, \tau_1)$. Let $T > t_0$ be fixed, and set $t = \min(\tau, T)$. Then for $i \in I_0$, we find that

$$E(m_i(t)) = E\left( \int_{t_0}^t g_i(s, m(s)) \, ds \right). \tag{4.5.2}$$

Since $t \le \tau$, we see that $g_i(s, m(s)) \ge 0$ for $s \in [t_0, t]$ and for each $i \in I_0$. Hence for $i \in I_0$, Lemma 4.5.2 gives

$$E(|m_i(t)|) \le E(|m_{0i}|) + E\left( \int_{t_0}^t g_i(s, m(s)) \, ds \right),$$

which, together with (4.5.2) and the fact that $m_{0i}(t) - 0$, yields $E(|m_i(t)|) \le E(m_i(t))$. This implies that for $i \in I_0$, $m_i(t) \ge 0$ a.s. From the definitions of $\tau$, $\tau_1$, and the nature of $m_{i0}$, $m_i(t) \ge 0$ a.s. for $i \in I \backslash I_0$. This implies that $m(t) \ge 0$ a.s. Since this is true for all $t = \min(\tau, T)$, the sample continuity of $m(t)$ shows that

$$P(t \in [t_0, \tau] \Rightarrow m(t) \ge 0) = 1,$$

which proves (4.5.1).

To prove that $m(t) \ge 0$ a.s. for all $t \ge t_0$, we let $t_1 = \inf \bigcup_{i=1}^n [s : m_i(s) < 0]$. It is enough to show that $t_1 = \infty$ a.s. Suppose that $P(t_1 < \infty) > 0$. Set $\tilde{\Omega} = [\omega : t_1(\omega) < \infty]$, $\tilde{\mathscr{F}}_t = \mathscr{F}_{t+t_1-t_0} | \tilde{\Omega}$, $\tilde{\mathscr{F}} = \mathscr{F} | \tilde{\Omega}$, and $\tilde{P}(A) = P(A) / P(\tilde{\Omega})$, $A \in \tilde{\mathscr{F}}$. On the space $(\tilde{\Omega}, \tilde{\mathscr{F}}, \tilde{P})$, we set $\tilde{G}(t, u) = G(t + t_1 - t_0, u)$, $\tilde{g}(t, u) = g(t + t_1 - t_0, u)$, $\tilde{m}(t) = m(t + t_1 - t_0)$, and $\tilde{z}(t) = z(t + t_1 - t_0)$. Then it follows that $\tilde{m}(t_0) = m(t_1) \in \partial R^n_+$ a.s. and

$$d\tilde{m} = \tilde{g}(t, \tilde{m}) \, dt + \tilde{G}(t, \tilde{m}) \, d\tilde{z}(t), \qquad \tilde{m}(t_0) = \tilde{m}_0.$$

We now apply (4.5.1) and obtain

$$\tilde{P}(\text{there exists a } t > t_0 : \tilde{m}(s) \ge 0 \text{ for } s \in [t_0, t]) = 1.$$

But this contradicts the definition of $t_1$. Hence $t_1 = \infty$ a.s., and hence $m(t) \ge 0$ a.s. for $t \ge t_0$, whenever $m_0 \in \partial R^n_+$.

To complete the proof of the theorem, we need to show that if $m_0 \in R_+^n \backslash \partial R_+^n$, then $m(t) \geq$ a.s. for $t \geq t_0$. To show this, set

$$t_1 = \inf \bigcup_{i=1}^{n} [s : m_i(s) < 0],$$

and by using an argument similar to the above, we can arrive at a contradiction to the definition of $t_1$. The proof of the theorem is complete.

**Corollary 4.5.1.** The conclusion of Theorem 4.5.1 remains valid when the strict quasi-positivity of $g$ for $t > t_0$ is weakened to quasinonnegativity for $t > t_0$ if the uniqueness of the solution is guaranteed for the Itô system.

*Proof.* Let $\varepsilon > 0$, and let $m(t, \varepsilon)$ be a solution of

$$dm = [g(t, m) + \varepsilon] dt + G(t, m) dz(t), \qquad m(t_0, \varepsilon) = m_0 + \varepsilon.$$

By Theorem 4.5.1, $m(t, \varepsilon) \geq 0$ a.s. for $t \geq t_0$. Now the uniqueness of the solutions implies that $m(t) = \lim_{\varepsilon \to 0} m(t, \varepsilon) \geq 0$ a.s. for $t \geq t_0$.

We shall next prove some basic results on differential inequalities.

**Theorem 4.5.2.** Assume that

(i)   $f \in C[R_+ \times R^n, R^n]$, $F \in C[R_+ \times R^n, R^{nm}]$, $z(t)$ is a normalized $m$-vector Wiener process, and $f(t, x)$ is quasi-monotone nondecreasing in $x$ for each $t$;

(ii)   $\sum_{j=1}^{m} |F_{ij}(t, u) - F_{ij}(t, v)|^2 \leq h_i(|u_i - v_i|)$ and $(t, u)$, $(t, v) \in R_+ \times R^n$, where for each $i = 1, 2, \ldots, n$, $h_i \in C[R_+, R_+]$, $h_i(0) = 0$, $h_i(s)$ is increasing in $s$, and $\int_{0+} ds/h_i(s) = \infty$;

(iii)   $u(t)$, $v(t)$ are solutions of

$$du = \beta_1(t) dt + F(t, u) dz(t), \qquad u(t_0) = u_0$$
$$dv = \beta_2(t) dt + F(t, v) dz(t), \qquad v(t_0) = v_0, \qquad t \geq t_0,$$

respectively, such that $\beta_1(t) \leq f(t, u(t))$ and $\beta_2(t) \geq f(t, v(t))$, one of the inequalities being strict.

Then $u_0 \leq v_0$ implies $u(t) \leq v(t)$, $t \geq t_0$, a.s.

*Proof.* We set $m(t) = v(t) - u(t)$ so that

$$dm = g(t, m) dt + G(t, m) dz(t), \tag{4.5.3}$$

where

$$G(t, m) = F(t, u + m) - F(t, u),$$
$$g(t, m) = f(t, u + m) - f(t, u) + q(t),$$
$$q(t) = \beta_2(t) - f(t, v) - \beta_1(t) + f(t, u).$$

From the quasi-monotonicity of $f(t,x)$ in $x$, it follows that $g(t,x)$ is quasi-monotone in $x$; moreover, if one of the inequalities $\beta_2 \geq f(t,v)$, $\beta_1 \leq f(t,u)$ is strict, we see that $g(t,x)$ is strictly quasi-positive in $x$ for $t > t_0$. In view of assumption (ii), $G(t,m)$ satisfies condition (ii) of Lemma 4.5.2. Hence by Theorem 4.5.1, we get $m(t) \geq 0$, $t \geq t_0$, a.s., which implies the stated result.

**Corollary 4.5.2.**  If in Theorem 4.5.2, the strictness of one of the inequalities $\beta_2 \geq f(t,v)$, $\beta_1 \leq f(t,u)$ is dropped and the uniqueness of solutions of (4.5.3) is assumed, the conclusion of Theorem 4.5.2 remains valid.

The following form of Theorem 4.5.2 is more suitable for later discussion. We merely state it.

**Theorem 4.5.3.**  Assume that

(i)  $u(t)$, $v(t)$ are solutions of

$$du = f_1(t,u)\,dt + F(t,u)\,dz(t), \qquad u(t_0) = u_0,$$
$$dv = f_2(t,v)\,dt + F(t,v)\,dz(t), \qquad v(t_0) = v_0, \qquad t \geq t_0,$$

respectively, where $f_1, f_2 \in C[R_+ \times R^n, R^n]$, $F \in C[R_+ \times R^n, R^{nm}]$, $z(t)$ is a normalized $m$-vector Wiener process, $f_2(t,x)$ is quasi-monotone non-decreasing in $x$ for each $t$, and $f_1(t,x) < f_2(t,x)$;

(ii)  condition (ii) of Theorem 4.5.2 holds.

Then $u_0 \leq v_0$ implies $u(t) < v(t)$, $t \geq t_0$, a.s.

It might seem that the results contained in Theorems 4.5.2 and 4.5.3 do not look like theorems on differential inequalities, although intrinsically they are. For example, in view of assumption (iii), it is clear that $u$, $v$ satisfy

$$\begin{aligned}
du &\leq f(t,u)\,dt + F(t,u)\,dz(t), \\
dv &\geq f(t,v)\,dt + F(t,v)\,dz(t),
\end{aligned} \tag{4.5.4}$$

and consequently Theorem 4.5.2 is implicitly a result on stochastic differential inequalities. Unfortunately, it is not possible to prove the assertion of Theorem 4.5.2 by starting from (4.5.4) as is usual.

## 4.6.  MAXIMAL AND MINIMAL SOLUTIONS

We shall introduce the notion of maximal and minimal solutions relative to the Itô-type stochastic system

$$du = f(t,u)\,dt + \sigma(t,u)\,dz(t), \qquad u(t_0) = u_0, \tag{4.6.1}$$

where $f \in C[J \times R^n, R^n]$, $\sigma \in C[J \times R^n, R^{nm}]$, and $z(t) \in R[\Omega, R^m]$ is a normalized Weiner process. Here $J = [t_0, t_0 + a)$. Let us begin by defining the extremal solutions.

**Definition 4.6.1.**    Let $r(t)$ be a solution of (4.6.1) on $J$. Then $r(t)$ is said to be the maximal solution of (4.6.1) if for every solution $u(t)$ existing on $J$, the inequality

$$u(t) \le r(t), \qquad t \in J, \quad \text{w.p. 1}, \tag{4.6.2}$$

holds. A minimal solution is defined similarly by reversing inequality (4.6.2).

We shall now prove the existence of extremal solutions.

**Theorem 4.6.1.**    Let assumptions $(H_1)$ and $(H_2)$ of Theorem 4.2.1 hold. Assume further that

$(H_3)$  (a)  $f(t, u)$ is quasi-monotone nondecreasing in $u$ for each $t$;
    (b)  $\sum_{j=1}^{m} |\sigma_{ij}(t, u) - \sigma_{ij}(t, v)|^2 \le h_i(|u_i - v_i|)$, $(t, u)$, $(t, v) \in J \times R^n$, where for each $i$, $h_i \in C[R_+, R_+]$, $h_i(0) = 0$, $h_i(s)$ is increasing in $s$, and $\int_{0+} ds/h_i(s) = \infty$.

Then there exist maximal and minimal solutions of (4.6.1) on $J$.

*Proof.*    We shall prove the existence of the maximal solution only since the case of the minimal solution is very similar.

Let $\varepsilon > 0$ be such that $\|\varepsilon\| \le L$, and consider the problem

$$du = [f(t, u) + \varepsilon] dt + \sigma(t, u) dz(t), \qquad u(t_0, \varepsilon) = u_0 + \varepsilon. \tag{4.6.3}$$

It is easy to see that the hypotheses of Theorem 4.2.1 are satisfied for (4.6.3) with $L$ replaced by $2L$. Hence there is a solution $u(t, \varepsilon)$ of (4.6.3) on $J$. Let $0 < \varepsilon_1 < \varepsilon_2 \le \varepsilon$, and let $u_1 = u(t, \varepsilon_1)$ $u_2 = u(t, \varepsilon_2)$ be solutions of

$$du_1 = [f(t, u_1) + \varepsilon_1] dt + \sigma(t, u_1) dz(t), \qquad u_1(t_0) = u_0 + \varepsilon_1,$$
$$du_2 = [f(t, u_2) + \varepsilon_2] dt + \sigma(t, u_2) dz(t), \qquad u_2(t_0) = u_0 + \varepsilon_2.$$

Then an application of Theorem 4.5.3 gives

$$u(t, \varepsilon_1) \le u(t, \varepsilon_2), \qquad t \in J, \quad \text{w.p. 1}.$$

Choose a decreasing sequence $\{\varepsilon_k\}$ such that $\varepsilon_k \to 0$ as $k \to \infty$. Then it is clear $\{u(t, \varepsilon_k)\}$ is a decreasing sequence and thus the uniform limit $r(t) = \lim_{k \to \infty} u(t, \varepsilon_k)$ w.p. 1 exists on $J$. Obviously $r(t_0) = u_0$. We have to show that $r(t)$ satisfies (4.6.1). For this purpose, we set

$$y(t) = u_0 + \int_{t_0}^{t} f(s, r(s)) ds + \int_{t_0}^{t} \sigma(s, r(s)) dz(s).$$

Let $J_0 = [t_0, t_0 + a_0] \subset J$. Then it follows that

$$f(t, u(t, \varepsilon_k)) \to f(t, r(t)),$$
$$\sigma(t, u(t, \varepsilon_k)) \to \sigma(t, r(t)),$$

as $k \to \infty$ uniformly on $J_0$. In view of this, after some manipulations (similar to the ones used in the last part of the proof of Theorem 4.2.1), we obtain

$$E(\|u(t, \varepsilon_k) - y(t)\|^2) \leq 3(t - t_0) E\left( \int_{t_0}^t \|f(s, u(s, \varepsilon_k)) - f(s, r(s))\|^2 \, ds \right)$$

$$+ 3 \int_{t_0}^t E(\|\sigma(s, u(s, \varepsilon_k)) - \sigma(s, r(s))\|^2) \, ds + 3\varepsilon_k^2$$

$$< \varepsilon^3 \qquad \text{for} \quad k \geq k_0, \quad t \in J_0.$$

Hence by Chebyshev's inequality, it follows that

$$P[\|u(t, \varepsilon_k) - y(t)\| > \varepsilon] < \varepsilon \qquad \text{on} \quad J_0 \qquad \text{for} \quad k \geq k_0,$$

which implies that

$$u(t, \varepsilon_k) \to y(t) \quad \text{w.p. 1} \qquad \text{as} \quad k \to \infty \qquad \text{on} \quad J_0.$$

This proves that $r(t)$ is a solution of (4.6.1) on $J$.

We shall now show that $r(t)$ is the desired maximal solution of (4.6.1). Let $u(t)$ be any solution of (4.6.1) on $J$. Then an application of Theorem 4.5.3 gives

$$u(t) \leq u(t, \varepsilon_k) \qquad \text{on} \quad J \quad \text{w.p. 1}.$$

This proves the theorem since $u(t) \leq \lim_{k \to \infty} u(t, \varepsilon_k) \equiv r(t)$ on $J$ w.p. 1.

## 4.7.   COMPARISON THEOREMS

An important technique in the theory of differential equations is concerned with estimating a function that satisfies a differential inequality by the extremal solutions of the corresponding differential equation. In this section we shall obtain comparison results of a similar type for the stochastic differential equation (4.6.1).

**Theorem 4.7.1.**   Assume that

(i)   $r(t)$ is the maximal solution of (4.6.1) on $J$;
(ii)   $m(t)$ is a solution of

$$dm = \beta(t) \, dt + \sigma(t, m) \, dz(t), \qquad m(t_0) = m_0, \qquad \text{on} \quad J,$$

where $\beta(t) \leq f(t, m(t))$ and $f(t, x)$ is quasi-monotone nondecreasing in $x$ for each $t$;
(iii)   condition $H_3(b)$ of Theorem 4.6.1 holds.

Then $m_0 \leq u_0$ implies $m(t) \leq r(t)$ on $J$ w.p. 1.

*Proof.*   Let $u(t, \varepsilon)$ be any solution of

$$du = [f(t, u) + \varepsilon] dt + \sigma(t, u) dz(t), \qquad u(t_0) = u_0 + \varepsilon,$$

for $\varepsilon > 0$ sufficiently small. Then by an application of Theorem 4.5.3, we have

$$m(t) \leq u(t, \varepsilon) \qquad \text{on} \quad J \quad \text{w.p. 1.}$$

Since $\lim_{\varepsilon \to 0} u(t, \varepsilon) \equiv r(t)$ on $J$, the proof is complete.

**Corollary 4.7.1.**   If in Theorem 4.7.1, the inequality $\beta(t) \leq f(t, m(t))$ is reversed, then the conclusion is to be replaced by

$$m(t) \geq \rho(t) \qquad \text{on} \quad J \quad \text{w.p. 1,}$$

where $\rho(t)$ is the minimal solution of (4.6.1).

The scalar version of Theorem 4.7.1, which we shall merely state below, is needed in later applications.

**Theorem 4.7.2.**   Assume that

(i)   $r(t)$ is the maximal solution of

$$du = f(t, u) dt + \sigma(t, u) dz(t), \qquad u(t_0) = u_0,$$

where $f, \sigma \in C[J \times R, R]$ and $z(t)$ is a normalized Weiner process;

(ii)   $m(t)$ is a solution of

$$dm = \beta(t) dt + \sigma(t, m) dz(t), \qquad m(t_0) = m_0,$$

where $\beta(t) \leq f(t, m(t))$;

(iii)   $|\sigma(t, u) - \sigma(t, v)|^2 \leq h(|u - v|)$ and $(t, u)$, $(t, v) \in J \times R$, where $h \in C[R_+, R_+]$, $h(0) = 0$, $h(t)$ is increasing in $s$, and $\int_{0^+} ds/h(s) = \infty$.

Then $m_0 \leq u_0$ implies $m(t) \leq r(t)$ on $J$ w.p. 1.

## 4.8.   LYAPUNOV-LIKE FUNCTIONS

As is well-known, the Lyapunov second method has played an important role in the qualitative study of solutions of differential equations. In this section, by using the concept of the Lyapunov function and the theory of differential inequalities (stochastic and deterministic), we develop some results that furnish very general comparison theorems.

**Definition 4.8.1.**   Let $G$ be a function on $R^n$ into $R^m$. The function $G$ is said to be convex if each component $G_i$ for $1 \leq i \leq m$ is convex, and $G$ is said to be concave if $-G_i$ is convex.

Consider the Itô-type stochastic differential system

$$dx = f(t, x)\, dt + \sigma(t, x)\, dz(t), \qquad x(t_0) = x_0, \qquad t_0 \in R_+, \qquad (4.8.1)$$

where $f \in C[R_+ \times R^n, R^n]$, $\sigma \in C[R_+ \times R^n, R^{nm}]$, and $z(t) \in R[\Omega, R^m]$ is a normalized Wiener process. We shall assume that the functions $f$ and $\sigma$ are smooth enough to assure the existence of the solution process.

Let $V \in C[R_+ \times R^n, R^N]$, $V_t$, $V_x$, and $V_{xx}$ exist and be continuous for $(t, x) \in R_+ \times R^n$. By Itô's formula, we obtain

$$dV(t, x) = LV(t, x)\, dt + V_x(t, x)\sigma(t, x)\, dz(t), \qquad (4.8.2)$$

where

$$LV(t, x) - V_t(t, x) + V_x(t, x)f(t, x) + \tfrac{1}{2} \operatorname{tr}(V_{xx}(t, x)\sigma(t, x)\sigma^{\mathrm{T}}(t, x)). \qquad (4.8.3)$$

We can now formulate the basic comparison theorems in terms of Lyapunov-like functions.

**Theorem 4.8.1.**   Assume that

(i)   $V \in C[R_+ \times R^n, R^N]$, $V_t$, $V_x$, and $V_{xx}$ exist and are continuous for $(t, x) \in R_+ \times R^n$, and for $(t, x) \in R_+ \times R^n$,

$$LV(t, x) \le g(t, V(t, x)), \qquad (4.8.4)$$

where $L$ is the operator defined in (4.8.3);

(ii)   $g \in C[R_+ \times R^N, R^N]$, $g(t, u)$ is concave and quasi-monotone non-decreasing in $u$ for each fixed $t \in R_+$, and $r(t) = r(t, t_0, u_0)$ is the maximal solution of the auxiliary differential system

$$u' = g(t, u), \qquad u(t_0) = u_0, \qquad (4.8.5)$$

existing for $t \ge t_0$;

(iii)   for the solution process $x(t) = x(t, t_0, x_0)$ of (4.8.1), $E[V(t, x(t))]$ exists for $t \ge t_0$.

Then

$$E[V(t, x(t))] \le r(t, t_0, u), \qquad t \ge t_0, \qquad (4.8.6)$$

whenever

$$E[V(t_0, x_0)] \le u_0. \qquad (4.8.7)$$

*Proof.*   Set

$$m(t) = E[V(t, x(t))] \qquad \text{for} \quad t \ge t_0.$$

Then assumption (iii), together with the continuity of $V(t, x)$ and the solution process $x(t)$, implies that $m(t)$ is continuous for $t \ge t_0$. Applying Itô's formula to $V(t, x(t))$, we get

$$dV = [V_t(t, x(t)) + V_x(t, x(t))f(t, x(t))$$
$$+ \tfrac{1}{2}\mathrm{tr}(V_{xx}(t, x(t))\sigma^{\mathrm{T}}(t, x(t))\sigma(t, x(t)))]\,dt$$
$$+ V_x(t, x(t))\sigma(t, x(t))\,dz(t).$$

For $h > 0$ sufficiently small, this, together with assumption (i) and the concavity of $g(t, u)$, implies

$$E[V(t + h, x(t + h))] - E[V(t, x(t))] = E\left[\int_t^{t+h} LV(s, x(s))\,ds\right]$$

$$\leq E\left[\int_t^{t+h} g(s, V(s, x(s)))\,ds\right]$$

$$\leq \int_t^{t+h} g(s, E[V(s, x(s))])\,ds. \qquad (4.8.8)$$

It therefore follows that

$$D^+ m(t) \leq g(t, m(t)), \qquad t \geq t_0. \qquad (4.8.9)$$

An application of Theorem 1.7.1 yields immediately the stated result (4.8.6), completing the proof.

**Corollary 4.8.1.** Suppose that conditions (i) and (ii) of Theorem 4.8.1 hold. Assume further that the initial value $u_0$ in (4.8.5) is a random variable so that Eq. (4.8.5) may be viewed as a random differential equation and that $E[V(t, x(t))|x_0]$ exists for $t \geq t_0$ w.p. 1. Then

$$E[V(t, x(t))|x_0] \leq r(t, t_0, u_0), \qquad t \geq t_0, \quad \text{w.p. 1},$$

provided that $V(t_0, x_0) \leq u_0$ w.p. 1. In particular, if $g(t, u) \equiv 0$, we have $E[V(t, x(t))|x(s)] \leq V(s, x(s)), t \geq s \geq t_0$, w.p. 1, which implies that $[V(t, x(t)): \mathscr{F}_t, t \geq t_0]$ is a supermartingale.

*Proof.* Proceeding as in the proof of Theorem 4.8.1 with $m(t) = E[V(t, x(t))|x_0]$ and noting that $m(t)$ is sample continuous, we arrive at the random differential inequality (4.8.9). We then apply the comparison theorem (Theorem 2.5.1) to derive the stated result. If $g \equiv 0$, we employ $m(t) = E[V(t, x(t))|x(s)], t \geq s \geq t_0$, to get the desired conclusion.

**Remark 4.8.1.** The drawback to Theorem 4.8.1 is assumption (iii). However, under certain conditions one could show that assumption (iii) holds. For example, let $V(t, x) \geq 0$ and $V(t, x) \leq a(t, \|x\|)$, where $a \in C[R_+ \times R_+, R_+^N]$ and $a(t, u)$ is concave in $u$ for fixed $t \in R_+$. Then we could have

$$0 \leq E[V(t, x(t))] \leq a(t, E[\|x(t)\|]), \qquad (4.8.10)$$

from which condition (iii) follows, using the concavity of $a$ and the fact that the solution process $x(t) \in L^2[\Omega, R^n]$.

Assumption (iii) in Theorem 4.8.1 can be avoided if we employ integral inequalities. But then other restrictions would be needed. Nonetheless, the comparison results that emerge from this approach are useful. We discuss such a comparison theorem next.

**Theorem 4.8.2.** Assume that hypotheses (i) and (ii) of Theorem 4.8.1 are satisfied except that the quasi-monotonic nondecreasing property of $g(t, u)$ in $u$ for fixed $t \in R_+$ is replaced by the nondecreasing property of $g(t, u)$ in $u$ for fixed $t \in R_+$. Suppose further that the function $V(t, x)$ is bounded below. Then the conclusion of Theorem 4.8.1 remains valid.

*Proof.* Let $x(t)$ be the solution process of the system (4.8.1) such that $E[V(t_0, x_0)] \leq u_0$. Let $\tau_n$ be the first exist time of $x(t)$ from $B(n) = \{x \in R^n : \|x\| \leq n\}$ for a positive integer $n \geq 1$. Define $\tau_n(t) = \min(t, \tau_n)$. Then $\tau_n(t)$ is a Markov time. By Itô's formula, we obtain

$$E[V(\tau_n(t), x(\tau_n(t)))] \leq E[V(t_0, x_0)] + E\left[\int_{t_0}^{\tau_n(t)} LV(s, x(s)) \, ds\right].$$

Hence condition (4.8.4) yields

$$E[V(\tau_n(t), x(\tau_n(t)))] \leq E[V(t_0, x_0)] + E\left[\int_{t_0}^{\tau_n(t)} g(s, V(s, x(s))) \, ds\right]. \quad (4.8.11)$$

This inequality, together with the continuity of the functions involved and the fact that $V(t, x)$ is bounded below, implies that $E[V(\tau_n(t), x(\tau_n(t)))]$ exists for all $t \geq t_0$. Clearly, for $t_0 \leq s \leq \tau_n(t)$, $m(s) = E[V(s, x(s))]$ is continuous. Furthermore, since $g(t, u)$ is concave in $u$, we have

$$E[g(s, V(s, x(s)))] \leq g(s, m(s)), \qquad t_0 \leq s \leq \tau_n(t). \quad (4.8.12)$$

Consequently, the assumption that $g(t, u) \geq 0$ and the continuity of $g(t, u)$ in $(t, u)$ yield

$$0 \leq \int_{t_0}^{s} E[g(\xi, V(\xi, x(\xi)))] \, d\xi \leq \int_{t_0}^{s} g(\xi, m(\xi)) \, d\xi < \infty$$

for $t_0 \leq s \leq \tau_n(t)$. Hence applying Fubini's theorem, we have

$$E\left[\int_{t_0}^{s} g(\xi, V(\xi, x(\xi))) \, d\xi\right] = \int_{t_0}^{s} E[g(\xi, V(\xi, x(\xi)))] \, d\xi.$$

This, together with (4.8.11), (4.8.12), and the definition of $m(s)$, leads to the integral inequality

$$m(t) \leq m(t_0) + \int_{t_0}^{s} g(\xi, m(\xi)) \, d\xi, \qquad t_0 \leq s \leq \tau_n(t).$$

We now apply Theorem 1.7.2 to obtain the inequality

$$E[V(s, x(s))] \leq r(s, t_0, u_0), \qquad t_0 \leq s \leq \tau_n(t),$$

which implies

$$E[V(\tau_n(t), x(\tau_n(t)))] \leq r(\tau_n(t), t_0, u_0).$$

Since $g(t, u) \geq 0$, the solutions $u(t, t_0, u_0)$ of (4.8.5) are nondecreasing in $t$, and consequently we have

$$E[V(\tau_n(t), x(\tau_n(t)))] \leq r(t, t_0, u_0).$$

We now apply Fatou's lemma to the left-hand side of this inequality to get the desired result

$$E[V(t, x(t))] \leq r(t, t_0, u_0), \qquad t \geq t_0.$$

The proof is complete.

The following variant of Theorem 4.8.1 is often more useful in applications.

**Theorem 4.8.3.**   Let the hypotheses of Theorem 4.8.1 hold except that inequality (4.8.4) is strengthened to

$$LA(t)V(t, x) \leq g(t, A(t)V(t, x)) \tag{4.8.13}$$

for $(t, x) \in R_+ \times R^n$, where $A(t)$ is a continuously differentiable $N \times N$ matrix function such that $A^{-1}(t)$ exists and its elements are nonnegative and continuous, and the elements of the matrix $A^{-1}(t)A'(t) = (\alpha_{ij}(t))$ satisfy $\alpha_{ij}(t) \leq 0$ for $i \neq j$. Then $E[A(t_0)V(t_0, x_0)] \leq u_0$ implies

$$E[V(t, x(t))] \leq R(t, t_0, u_0), \qquad t \geq t_0, \tag{4.8.14}$$

where $R(t, t_0, u_0)$ is the maximal solution of the auxiliary differential system

$$v' = A^{-1}(t)[-A'(t)v + g(t, A(t)v)], \qquad v(t_0) = v_0, \tag{4.8.15}$$

existing for $t \geq t_0$.

*Proof.*   Setting $W(t, x) = A(t)V(t, x)$, we see (because of (4.8.13)) that

$$LW(t, x) \leq g(t, W(t, x)).$$

This shows that the function $W(t, x)$ satisfies all the assumptions of Theorem 4.8.1, and as a consequence, we have

$$E[W(t, x(t))] \leq r(t, t_0, u_0), \qquad t \geq t_0, \tag{4.8.16}$$

provided that $E[W(t_0, x_0)] \leq u_0$. Here $r(t, t_0, u_0)$ is the maximal solution of (4.8.5). It is easy to see that

$$A(t)R(t, t_0, v_0) = r(t, t_0, u_0) \tag{4.8.17}$$

with $A(t_0)u_0 = v_0$. From (4.8.16), (4.8.17), the properties of the mean $E$, the definition of $W(t, x)$, and the properties of $A(t)$, we have

$$E[V(t, x(t))] \leq R(t, t_0, v_0), \qquad t \geq t_0.$$

Thus the proof is complete.

**Remark 4.8.2.**   Let $B(t)$ be a diagonal matrix function whose elements are nonnegative. Then the function $A(t) = \exp[\int_0^t B(s)\,ds]$ is admissible in Theorem 4.8.3.

In the following we shall present a basic comparison theorem in the framework of stochastic differential inequalities.

**Theorem 4.8.4.**   Assume that

(i)   $V \in C[R_+ \times R^n, R^N]$, $V_t$, $V_x$, and $V_{xx}$ exist and are continuous for $(t, x) \in R_+ \times R^n$ and that for $(t, x) \in R_+ \times R^n$,

$$dV(t, x) = \alpha(t, x)\,dt + G(t, V(t, x))\,dz(t), \tag{4.8.18}$$

where $\alpha \in C[R_+ \times R^n, R^N]$, $g \in C[R_+ \times R^N, R^N]$, $G \in C[R_+ \times R^N, R^{Nm}]$, $\alpha(t, x) \leq g(t, V(t, x))$, and $\alpha(t, x) = LV(t, x)$;

(ii)   (a)   $g(t, u)$ is quasi-monotone nondecreasing in $u$ for each $t$;
  (b)   $\sum_{j=1}^m |G_{ij}(t, u) - G_{ij}(t, v)|^2 \leq h_i(|u_i - v_i|)$, $(t, u)$, $(t, v) \in R_+ \times R^N$, where $h_i \in C[R_+, R_+]$, $h_i(0) = 0$, $h_i(s)$ is increasing in $s$, and $\int_{0+} ds/h_i(s) = \infty$, $i = 1, 2, \ldots, N$;

(iii)   $r(t) = r(t, t_0, u_0)\cdot$ is the maximal solution of the Itô-type stochastic differential system

$$du = g(t, u)\,dt + G(t, u)\,dz(t), \tag{4.8.19}$$

existing for $t \geq t_0$.

(iv)   $x(t) = x(t, t_0, x_0)$ is a solution process of (4.8.1) such that

$$V(t_0, x_0) \leq u_0 \quad \text{w.p. 1.} \tag{4.8.20}$$

Then

$$V(t, x(t)) \leq r(t, t_0, u_0), \qquad t \geq t_0, \quad \text{w.p. 1.} \tag{4.8.21}$$

*Proof.*   Let $x(t)$ be any solution process of (4.8.1) defined for $t \geq t_0$ such that (4.8.20) holds. Define

$$m(t) = V(t, x(t))$$

so that $m(t_0) \leq u_0$. This, together with (4.8.18), yields

$$dm(t) = \beta(t)\,dt + G(t, m(t))\,dz(t),$$

where $\beta(t) \equiv \alpha(t, x(t)) = LV(t, x(t))$. Applying Theorem 4.7.1, we obtain the desired result (4.8.21).

## 4.9. STABILITY IN PROBABILITY

Let $x(t) = x(t, t_0, x_0)$ be any solution of the Itô-type stochastic differential system (4.8.1). Without loss of generality, we assume that $f(t, 0) \equiv 0$ and $\sigma(t, 0) \equiv 0$ for all $t \in R_+$ and that $x(t) \equiv 0$ is the unique solution of (4.8.1) through $(t_0, 0)$. The stability concepts $(SP_1)$–$(SP_4)$, $(SS_1)$–$(SS_4)$, and $(SM_1)$–$(SM_4)$ of Section 2.10 with respect to the trivial solution $x \equiv 0$ of (4.8.1) will be used in our discussion.

To derive the stability properties of the trivial solution of (4.8.1) in the framework of the second method of Lyapunov, we need to know the stability properties of the corresponding auxiliary or comparison differential system. We shall utilize the stochastic comparison system (4.8.19) as well as the deterministic system (4.8.5), where $g$, $G$, and $z$ are as defined in Theorem 4.8.4.

Corresponding to the stability definitions $(SP_1)$–$(SP_4)$, $(SS_1)$–$(SS_4)$, and $(SM_1)$–$(SM_4)$, we shall designate by $(SP_1^*)$–$(SP_4^*)$, $(SS_1^*)$–$(SM_4^*)$, and $(SM_1^*)$–$(SM_4^*)$ the concepts of Section 2.10 concerning the stability of the equilibrium solution $u \equiv 0$ of (4.8.19). Similarly, we shall denote by $(S_1^*)$–$(S_4^*)$ the concepts of Section 3.6 concerning the stability of the equilibrium solution of $u \equiv 0$ of (4.8.5).

We begin with the following stability criteria, recalling the definition of the function $LV(t, x)$ given in (4.8.3).

**Theorem 4.9.1.**    Assume that

(i)   $g \in C[R_+ \times R_+^N, R^N]$, $g(t, 0) \equiv 0$, $g(t, u)$ is concave and quasi-monotone nondecreasing in $u$ for each $t \in R_+$;

(ii)   $V \in C[R_+ \times B(\rho), R^N]$, $V_t(t, x)$, $V_x(t, x)$, and $V_{xx}(t, x)$ exist and are continuous on $R_+ \times B(\rho)$, and for $(t, x) \in R_+ \times B(\rho)$,

$$LV(t, x) \leq g(t, V(t, x)),$$

where $B(\rho) = [x \in R^n : \|x\| < \rho]$;

(iii)   for $(t, x) \in R_+ \times B(\rho)$,

$$b(\|x\|) \leq \sum_{i=1}^{N} V_i(t, x) \leq a(t, \|x\|), \tag{4.9.1}$$

where $b \in \mathscr{K}$, $a \in \mathscr{CK}$.

Then

$$(S_1^*) \qquad \text{implies} \qquad (SP_1).$$

*Proof.*   Let $x(t)$ be the solution process associated with (4.8.1). By assumption (iii), we have

$$0 \leq E[b(\|x(t)\|)] \leq E\left[\sum_{i=1}^{n} V_i(t, x(t))\right] \leq a(t, E[\|x(t)\|])$$

so long as $x(t) \in B(\rho)$ for $t \geq t_0$. This inequality assures, as noted in Remark 4.8.1, that $E[V(t, x(t))]$ exists. Hence by Theorem 4.8.1, the estimate

$$E[V(t, x(t))] \leq r(t, t_0, u_0) \qquad (4.9.2)$$

is valid so long as $x(t) \in S(\rho)$ for $t \geq t_0$, provided that

$$E[V(t_0, x_0)] \leq u_0. \qquad (4.9.3)$$

It is obvious that the relation (4.9.2) yields the estimate

$$\sum_{i=1}^{N} E[V_i(t, x(t))] \leq \sum_{i=1}^{N} r_i(t, t_0, u_0). \qquad (4.9.4)$$

Let $0 < \eta < 1, 0 < \varepsilon < \rho$, and $t_0 \in R_+$ be given. Then given $\varepsilon_1 = \eta b(\varepsilon) > 0$ and $t_0 \in R_+$, there exists a positive function $\delta_1 = \delta_1(t_0, \varepsilon, \eta)$ that is continuous in $t_0$ for each $\varepsilon$ and $\eta$ such that $\sum_{i=1}^{N} u_{i0} \leq \delta_1$ implies

$$\sum_{i=1}^{N} u_i(t, t_0, u_0) < \eta b(\varepsilon), \qquad t \geq t_0, \qquad (4.9.5)$$

where $u(t, t_0, u_0)$ is any solution of (4.8.5). We choose $u_0$ so that

$$\sum_{i=1}^{N} u_{i0} = a(t_0, E[\|x_0\|]). \qquad (4.9.6)$$

Since $a \in \mathscr{CK}$, we can find a $\delta_2 = \delta_2(t_0, \varepsilon, \eta) > 0$ that is continuous in $t_0$ for each $\varepsilon$ and $\eta$ such that the inequalities

$$E[\|x_0\|] \leq \delta_2 \qquad \text{and} \qquad a(t_0, E[\|x_0\|]) \leq \delta_1 \qquad (4.9.7)$$

hold simultaneously. Now we choose a $\delta = \delta(t_0, \varepsilon, \eta)$ that is continuous in $t_0$ for each $\varepsilon$ and $\eta$ such that $\delta_2 < \eta\delta$. This, together with Chebyshev's inequality and (4.9.7), yields

$$P[\omega : \|x_0(\omega)\| > \delta] < \eta. \qquad (4.9.8)$$

We claim that $(SP_1)$ holds with this $\delta$. If this claim is not true, then there exists a $t_1 > t_0$ such that

$$P[\|x(t_1)\| \geq \varepsilon] = \eta. \qquad (4.9.9)$$

Let $\tau_\varepsilon$ be the first exist time of $x(t)$ from $\{x \in R^n : \|x\| < \varepsilon\}$, and let $\tau_\varepsilon(t) = \min(t, \tau_\varepsilon)$. Hence from (4.9.4) and (4.9.5), we get

$$\sum_{i=1}^{N} E[V_i(\tau_\varepsilon(t), x(\tau_\varepsilon(t)))] < \eta b(\varepsilon),$$

which implies that

$$\sum_{i=1}^{N} \left[ \int_{\Omega_1} V_i(\tau_\varepsilon(t), x(\tau_\varepsilon(t))) P(d\omega) + \int_{\Omega_2} V_i(\tau_\varepsilon(t), x(\tau_\varepsilon(t))) P(d\omega) \right] < \eta b(\varepsilon),$$

where $\Omega_1 = [\omega \in \Omega : \tau_\varepsilon(t) > t_1]$ and $\Omega_2 = [\omega \in \Omega : \tau_\varepsilon(t) \le t_1]$. This, in view of condition (iii), gives

$$b(\varepsilon)P[\tau_\varepsilon \le t_1] \le \sum_{i=1}^{N} \int_\Omega V_i(\tau_\varepsilon(t), x(\tau_\varepsilon(t)))P(d\omega) < \eta b(\varepsilon).$$

Since

$$P\left[\sup_{t_0 \le t < t_1} \|x(t)\| \ge \varepsilon\right] = P[\tau_\varepsilon(t) \le t_1],$$

the last inequality leads to

$$P[\|x(t_1)\| \ge \varepsilon] < \eta,$$

which contradicts relation (4.9.9). This shows that $(SP_1)$ is true.

**Theorem 4.9.2.**   Let the hypotheses of Theorem 4.9.1 hold with $a(t,r) = a(r)$. Then

$$(S_2^*) \qquad \text{implies} \qquad (SP_2).$$

*Proof.*   By following the proof of Theorem 4.9.1, it is easy to see that $\delta$ does not depend on $t_0$. For by the assumptions of the uniform stability of the null solution $u = 0$ of (4.8.5) and $a(t,r) = a(r)$, $\delta_1$ and $\delta_2$ are independent of $t_0$. This shows that $\delta$ is also independent of $t_0$.

**Corollary 4.9.1.**   The functions $g(t,u) = \lambda(t)u$, where $\int_0^t \lambda(s)\,ds$ and $g(t,u) = 0$ are admissible in Theorem 4.9.2.

**Theorem 4.9.3.**   Let the hypotheses of Theorem 4.9.1 be satisfied. Then

$$(S_3^*) \qquad \text{implies} \qquad (SP_3).$$

*Proof.*   We now fix $\varepsilon = \rho$ and $\eta = \eta_0$ and designate by $\delta$ the number $\delta(t_0)$. To prove the conclusion, assume that the trivial solution of (4.8.5) is equiasymptotically stable. In view of the proof of Theorem 4.9.1, it is enough to show that the trivial solution of (4.8.1) satisfies latter part of $(SP_3)$ for every $0 < \eta < \eta_0$, $0 < \varepsilon < \rho$. For this purpose, let $0 < \eta < \eta_0$, $0 < \varepsilon < \rho$, and $t_0 \in R_+$ be given. By the equiasymptotic stability of $u \equiv 0$ of (4.8.5), given $\eta b(\varepsilon)$ and $t_0 \in R_+$, there exist numbers $\delta_1^0 = \delta_1^0(t_0) > 0$ and $T = T(t_0, \varepsilon, \eta) > 0$ such that

$$\sum_{i=1}^{N} u_i(t, t_0, u_0) < \eta b(\varepsilon), \qquad t \ge t_0 + T, \qquad (4.9.10)$$

whenever $\sum_{i=1}^{N} u_{i0} \le \delta_1^0$. As before, we choose $u_0$ so that (4.9.9) holds and also choose $\delta_2^0(t_0)$ so that (4.9.7) holds. Choose $\delta_0$ so that $\min(\delta, \delta_2^0) < \eta\delta_0$.

This, together with Chebyshev's inequality and (4.9.7), gives

$$P[\omega : \|x_0(\omega)\| > \delta_0] < \eta, \tag{4.9.11}$$

which implies that

$$\sum_{i=1}^{N} E[V_i(t, x(t))] \le \sum_{i=1}^{N} r_i(t, t_0, u_0), \qquad t \ge t_0,$$

with probability greater than or equal to $1 - \eta_0$, since $\|x(t)\| < \varepsilon$ with probability greater than or equal to $1 - \eta_0$. As a result, we have (because of (4.9.10))

$$\sum_{i=1}^{N} E[V_i(t, x(t))] < \eta b(\varepsilon), \qquad t \ge t_0 + T,$$

with probability greater than or equal to $1 - \eta_0$, which implies, arguing as before, that

$$P[\|x(t)\| \ge \varepsilon] < \eta, \qquad t \ge t_0 + T,$$

whenever (4.9.11) holds. This proves ($SP_3$), and therefore the proof is complete.

**Theorem 4.9.4.** Assume that all the hypotheses of Theorem 4.9.2 hold. Then

$$(S_4^*) \qquad \text{implies} \qquad (SP_4).$$

*Proof.* The proof follows from the proofs of Theorems 4.9.2 and 4.9.4.

**Remark 4.9.1.** Instead of using the comparison theorem (Theorem 4.8.1) for the stability considerations as was done in the foregoing discussion, we could utilize the comparison result contained in Corollary 4.8.1. This would necessitate employing the random differential system (4.8.5) (note that only the initial value will be random) as the comparison system. Thus if the trivial solution of (4.8.5) is stable in probability ($SP_1^*$), we would then obtain the following type of stability for the given system (4.8.1), namely, $P[\|x_0\| \ge \delta] < \eta$ implies $P[E(\|x(t)\|x_0\|) \ge \varepsilon] < \eta$, $t \ge t_0$. On the other hand, if we assume stability in the mean for the comparison system, we have a similar concept holding for (4.8.1), provided that the function $b(u)$ is also convex (refer to Theorem 4.9.2). In particular, if the initial value $x_0$ is constant w.p. 1 and $g \equiv 0$, we get stability in probability for (4.8.1). Thus it is clear that the comparison result in Corollary 4.8.1 offers another unified and systematic approach to investigate slightly different kinds of stability notions. We do not wish to pursue this treatment.

As an illustration consider the Itô-type linear system of stochastic differential equations

$$dx = Ax\,dt + \sum_{i=1}^{m} B_i x\,dz_i, \qquad (4.9.12)$$

where $z(t)$ is a Wiener process and $A$ and $B_i$, $i = 1, 2, \ldots, m$, are $n \times n$ constant matrices. Take $V(x) = x^{\mathrm{T}} Px$ as a Lyapunov function, where $P$ is the positive definite matrix solution of

$$0 = Q + PA + A^{\mathrm{T}}P + \sum_{i=1}^{m} B_i^{\mathrm{T}} PB_i \qquad (4.9.13)$$

and $Q$ is any given positive definite matrix. Noting that $g(t, u) = -\alpha u$, where $\alpha = (\lambda_m(P^{-1}Q))$ and $\lambda_m(P^{-1}Q)$ is the smallest eigenvalue of the matrix $P^{-1}Q$, it is clear that the trivial solution of (4.9.12) is uniformly asymptotically stable in probability.

The following example demonstrates the advantage of a vector Lyapunov function over a single Lyanpunov function.

**Example 4.9.1.** Consider the system of stochastic differential equations

$$dx = F(t, x)x\,dt + \sigma(t, x)\,dz, \qquad (4.9.14)$$

where $x \in R^2$, $z(t)$ is a normalized scalar Wiener process, $\sigma \in C[R_+ \times R^2, R^2]$, $\sigma(t, 0) \equiv 0$,

$$[\sigma_1(t, x) + \sigma_2(t, x)]^2 \le (x_1 + y_2)^2 \lambda(t),$$
$$[\sigma_1(t, x) - \sigma_2(t, x)]^2 \le (x_1 - x_2)^2 \lambda(t), \qquad (4.9.15)$$

$\lambda \in C[R_+, R_+] \cap L_1[0, \infty)$, and

$$F(t, x) = \begin{bmatrix} e^{-t} - f_1(t, x) & \sin t \\ \sin t & e^{-t} - f_1(t, x) \end{bmatrix},$$

$f_1 \in C[R_+ \times R^2, R]$, $f_1(t, x) \ge 0$ on $B(\rho)$, and $f_1(t, 0) \equiv 0$.

First, we choose a single Lyapunov function $V(t, x)$, given by $V(t, x) = x_1^2 + x_2^2$. Then it is evident that

$$LV(t, x) \le [2e^{-t} + 2|\sin t| + \lambda(t)]V(t, x),$$

using the inequalities $2|ab| \le a^2 + b^2$ and $f_1(t, x) \ge 0$ in $B(\rho)$. Clearly, the trivial solution of the comparison equation $u' = [2e^{-t} + 2|\sin t| + \lambda(t)]u$ is not stable. Hence we can not deduce any information about the stability of (4.9.14) from the scalar version of Theorem 4.9.1, that is, with $N = 1$, even though it is stable.

Now we attempt to seek stability information of (4.9.14) by employing a vector Lyapunov function. We choose

$$V(t, x) = \begin{bmatrix} (x_1 + y_2)^2 \\ (x_1 - y_2)^2 \end{bmatrix}.$$

Note that the components of $V$, $V_1$, and $V_2$ are not positive definite. However, $V_1$ and $V_2$ satisfy the hypotheses of Theorem 4.9.1 with $N = 2$. In fact,

$$(x_1^2 + x_2^2) \leq \sum_{i=1}^{2} V_i(t, x) \leq 2(x_1^2 + y_2^2)$$

and the vectorial inequality

$$LV(t, x) \leq g(t, V(t, x))$$

are satisfied in $B(\rho)$ with

$$g(t, u) = \begin{bmatrix} (2e^{-t} + 2\sin t + \lambda(t))u_1 \\ (2e^{-t} + 2\sin t + \lambda(t))u_2 \end{bmatrix}.$$

It is easy to observe that $g(t, u)$ is concave and quasi-monotone nondecreasing in $u$ for fixed $t$ and that the trivial solution of (4.8.5) is uniformly stable. Consequently, the trivial solution of (4.9.14) is stable in probability.

In the following, we shall illustrate the use of the variation of parameters method to study the stability in probability of Itô-type stochastic differential systems.

**Theorem 4.9.5.** Assume that

(i)   $\sigma(t, x)$ in (4.8.1) is an $n \times 1$ matrix function such that $\sigma(t, x) \equiv B(t)x$ and $z$ is a scalar normalized Wiener process, where $B \in C[R_+, R^{n^2}]$ is an $n \times n$ matrix function;

(ii)   for any $\eta > 0$ and $t_0 \in R_+$, there exists a positive number $T = T(t_0, \eta)$ such that

$$P\{\omega : \beta(t, t_0) \geq 0\} < \eta, \qquad t \geq t_0 + T,$$

where $\beta(t, t_0)$ is the largest eigenvalue of the stochastic matrix function

$$-\tfrac{1}{2}\int_{t_0}^{t} B^{T}(u)B(u)\, du + \int_{t_0}^{t} B(u)\, dz(u);$$

(iii)   $f(t, 0) \equiv 0$ for $t \in R_+$;

(iv)   $\|f(t, x)\| = o(\|x\|)$ as $x \to 0$ uniformly in $t \in R_+$.

Then the trivial solution of (4.8.1) is asymptotically stable in probability.

$f$.   From Theorem 4.4.2, we have

$$y(t, t_0, x_0) = \phi(t, t_0)\left[ x_0 + \int_{t_0}^{t} \phi^{-1}(s, t_0) f(s, y(s, t_0, x_0)) \, ds \right], \quad (4.9.16)$$

e

$$\phi(t, t_0) = \exp\left[ -\tfrac{1}{2} \int_{t_0}^{t} B^{T}(u)B(u) \, du + \int_{t_0}^{t} B(u) \, dz(u) \right]$$

he fundamental matrix solution process of

$$dx = B(t)x \, dz. \tag{4.9.17}$$

Let $\varepsilon > 0$ be sufficiently small. Then because of hypothesis (iv), there exists a $\delta > 0$ such that $\|x\| < \delta$ implies $\|f(t, x)\| < \varepsilon\|x\|$. This, together with (4.9.16) and the argument that is used in Theorem 2.11.7, yields

$$\|y(t, t_0, x_0)\| \le K_1 \|x_0\| \exp[\varepsilon(t - t_0) + \beta(t, t_0)] \tag{4.9.18}$$

so long as $\|y(t, t_0, x_0)\| < \delta$ w.p. 1, where $K_1$ satisfies

$$\left\| \exp\left[ -\tfrac{1}{2} \int_{t_0}^{t} B^{T}(u)B(u) \, du + \int_{t_0}^{t} B(u) \, dz(u) \right] \right\| \le K_1 \exp[\beta(t, t_0)]$$

With the foregoing $\varepsilon$ and hypothesis (ii), one can find a positive constant $K$ such that

$$P\{\omega : \varepsilon(t - t_0) + \beta(t, t_0) \ge K\}, < \eta \qquad t \ge t_0 + T,$$

and choose $\delta = \varepsilon/K_1 e^{K}$. Now the remaining proof of the theorem can be reformulated on the basis of the proof of Theorem 2.11.7.

**Problem 4.9.1.**   Assume that hypotheses (i), (iii), and (iv) of Theorem 4.9.5 hold. Further assume that for any $\eta > 0$ and $t_0 \in R_+$, there exists a positive number $T = T(t_0, \eta)$ such that

$$P\left\{ \omega : \int_{t_0}^{t} b(s) \, ds + \left\| \int_{t_0}^{t} B(s) \, dz(s) \right\| \ge 0 \right\} < \eta, \qquad t \ge t_0 + T,$$

where $b(t)$ is the largest eigenvalue of $-\tfrac{1}{2}B^{T}(t)B(t)$. Show that the trivial solution of (4.8.1) is asymptotically stable in probability.

### 4.10.   STABILITY IN THE $P$TH MEAN

In this section we shall present some stability criteria that assure the $p$th mean stability properties of the trivial solution of (4.8.1).

**Theorem 4.10.1.** Assume that

(i) $g \in C[R_+ \times R_+^N, R^N]$, $g(t,0) \equiv 0$, and $g(t,u)$ is concave and quasi-monotone nondecreasing in $u$ for $t \in R_+$;

(ii) $V \in C[R_+ \times R^n, R^N]$, $V_t(t,x)$, $V_x(t,x)$, and $V_{xx}(t,x)$ exist and are continuous on $R_+ \times R^n$, and for $(t,x) \in R_+ \times R^n$,

$$LV(t,x) \leq g(t, V(t,x));$$

(iii) for $(t,x) \in R_+ \times R^n$,

$$b(\|x\|^p) \leq \sum_{i=1}^{N} V_i(t,x) \leq a(t, \|x\|^p),$$

where $b \in \mathcal{VK}$, $a \in \mathcal{CK}$, and $p \geq 1$;

Then

$$(S_1^*) \quad \text{implies} \quad (SM_1).$$

*Proof.* By Theorem 4.8.1, we have

$$E[V(t, x(t))] \leq r(t, t_0, u_0), \qquad t \geq t_0, \tag{4.10.1}$$

for the solution process $x(t) = x(t, t_0, x_0)$ of (4.8.1) whenever

$$E[V(t_0, x_0)] \leq u_0, \tag{4.10.2}$$

where $r(t, t_0, u_0)$ is the maximal solution of (4.8.5). Let $\varepsilon > 0$ and $t_0 \in R_+$ be given, and suppose that $u \equiv 0$ of (4.8.5) is equistable. Then given $b(\varepsilon^p) > 0$ and $t_0 \in R_+$, there exists a $\delta_1 = \delta_1(t_0, \varepsilon)$ such that $\sum_{i=1}^N u_{i0} \leq \delta_1$ implies

$$\sum_{i=1}^{N} u_i(t, t_0, u_0) < b(\varepsilon^p), \qquad t \geq t_0, \tag{4.10.3}$$

where $u(t, t_0, u_0)$ is any solution of (4.8.5). Let us choose $u_0$ so that

$$E[V(t_0, x_0)] \leq u_0 \quad \text{and} \quad \sum_{i=1}^{N} u_{i0} = a(t_0, E[\|x_0\|^p]). \tag{4.10.4}$$

Since $a \in \mathcal{CK}$, we can find a $\delta = \delta(t_0, \varepsilon) > 0$ that is continuous in $t_0$ for each $\varepsilon > 0$ such that

$$\|x_0\|_p \leq \delta \quad \text{implies} \quad a(t_0, E[\|x_0\|^p]) < \delta_1.$$

Now we claim that $(SM_1)$ holds. Suppose that this claim is false. Then there would exist a solution process $x(t)$ of (4.8.1) with $\|x_0\|_p \leq \delta$ and $t_1 > t_0$ such that

$$\|x(t_1, t_0, x_0)\|_p = \varepsilon. \tag{4.10.5}$$

From (iii), we have

$$b(E[\|x(t)\|^p]) \leq \sum_{i=1}^{N} E[V_i(t, x(t))], \qquad t \geq t_0. \tag{4.10.6}$$

Relations (4.10.1), (4.10.3), (4.10.5), and (4.10.6) lead us to the contradiction

$$b(\varepsilon^p) \leq \sum_{i=1}^{N} E[V_i(t_1, x(t_1))] \leq \sum_{i=1}^{N} u_i(t_1, t_0, x_0) < b(\varepsilon^p),$$

thus proving the theorem.

Based on Theorem 4.10.1, it is easy to state and prove other types of $p$th mean stability properties. In order to avoid monotony, we do not venture proving these results.

In the following we give some results that show the role of the method of variation of parameters in the stability analysis of Itô-type systems. First we consider a stability result concerning (4.8.1) as a perturbation of (4.4.5).

**Theorem 4.10.2.** Assume that hypotheses (i), (iii), and (iv) of Theorem 4.9.5 are satisfied. Further assume that there exists a positive number $a = a(t_0)$ and a function $\lambda \in C[R_+, R_+]$ such that

$$-a(t_0) \equiv \limsup_{t \to \infty} \left[ \frac{p}{(t - t_0)} \left( \int_{t_0}^{t} [b(s) + \lambda(s)] \, ds \right) \right] \tag{4.10.7}$$

and

$$E\left[ \exp\left[ p \left\| \int_{t_0}^{t} B(u) \, dz(u) \right\| \right] \right] \leq \exp\left[ p \int_{t_0}^{t} \lambda(s) \, ds \right], \tag{4.10.8}$$

where $b(t)$ is the largest eigenvalue of $-\frac{1}{2}B^{\mathrm{T}}(t)B(t)$. Then the trivial solution of (4.8.1) is $p$th mean asymptotically stable.

*Proof.* From (4.9.16), we get

$$\|y(t, t_0, x_0)\| \leq K\left[ \|x_0\| \exp\left[ \int_{t_0}^{t} b(s) \, ds + \left\| \int_{t_0}^{t} B(s) \, dz(s) \right\| \right] \right.$$

$$+ \int_{t_0}^{t} \left( \exp\left[ \int_{s}^{t} b(u) \, du + \left\| \int_{s}^{t} B(u) \, dz(u) \right\| \right] \right.$$

$$\left. \times \|f(s, y(s, t_0, x_0))\| \right) ds \bigg].$$

This, together with hypothesis (iv), yields

$$\|y(t, t_0, x_0)\| \leq K\|x_0\| \exp\left[ \varepsilon(t - t_0) + \int_{t_0}^{t} b(s) \, ds + \left\| \int_{t_0}^{t} B(u) \, dz(u) \right\| \right] \tag{4.10.9}$$

so long as $\|y(t, t_0, x_0)\| < \delta$ w.p. 1, where $\delta = \varepsilon/K$. First, we take the $p$th exponent and then the expected value of the inequality, and using the fact that $x_0$ is independent of the increment $z(t) - z(s)$ of the Wiener process $z(t)$ for $t \geq s$, we get

$$E[\|y(t, t_0, x_0)\|^p] \leq \|x_0\|^p \exp\left[ p\left( \varepsilon(t - t_0) + \int_{t_0}^t b(s)\, ds \right) \right]$$

$$\times E\left[ \exp\left[ p \left\| \int_{t_0}^t B(u)\, dz(u) \right\| \right] \right]$$

so long as $\|y(t, t_0, x_0)\| < \delta$ w.p. 1. This, together with (4.10.8), yields

$$E[\|y(t, t_0, x_0)\|^p] \leq \|x_0\| \exp\left[ p\left( \varepsilon(t - t_0) + \int_{t_0}^t [b(s) + \lambda(s)]\, ds \right) \right] \qquad (4.10.10)$$

so long as $\|y(t, t_0, x_0)\| < \delta$ w.p. 1. Now by using relation (4.10.7) and the argument of Theorem 2.11.7, one can conclude that the trivial solution of (4.8.1) is asymptotically stable in the $p$th mean.

To derive the $p$th mean stability properties of (4.8.1), one can also treat (4.8.1) as a perturbation of (4.4.1). In this case one can use the variation of constants formula (4.4.3) to obtain the $p$th moment stability properties of (4.8.1). For more details, we refer the reader to Kulkarni and Ladde [47].

## 4.11.  STABILITY WITH PROBABILITY ONE

In this section we shall give some a.s. sample stability results concerning the trivial solution of (4.8.1). Here we shall utilize the stochastic comparison system (4.8.19). Some a.s. stability results will also be formulated by means of the variation of constants formula.

First, we shall present results in the framework of Lyapunov-like functions and Itô-type stochastic differential inequalities.

**Theorem 4.11.1.**   Assume that

(i)   hypothesis (ii) of Theorem 4.8.4 holds;
(ii)   $V \in C[R_+ \times B(\rho), R^N]$, $V_t(t, x)$, $V_x(t, x)$, and $V_{xx}(t, x)$ exist and are continuous for $(t, x) \in R_+ \times B(\rho)$, and for $(t, x) \in R_+ \times B(\rho)$,

$$dV(t, x) = \alpha(t, x)\, dt + G(t, V(t, x))\, dz(t),$$

where $\alpha \in C[R_+ \times B(\rho), R^N]$, $g \in C[R_+ \times R^N, R^N]$, $G \in C[R_+ \times R^N, R^{Nm}]$, $\alpha(t, x) \leq g(t, V(t, x))$, and $\alpha(t, x) = LV(t, x)$;

(iii)   for $(t, x) \in R_+ \times B(\rho)$,

$$b(\|x\|) \leq \sum_{i=1}^{N} V_i(t, x) \leq a(t, \|x\|),$$

where $a(t, \cdot)$ and $b(\cdot) \in \mathcal{K}$.

Then
$$(\text{SS}_1^*) \qquad \text{implies} \quad (\text{SS}_1).$$

*Proof.* Let $x(t)$ be a solution process associated with (4.8.1). By the application of Theorem 4.8.4, we have

$$V(t, x(t)) \leq r(t, t_0, u_0) \quad \text{w.p. 1} \qquad (4.11.1)$$

so long as $x(t) \in S(\rho)$ for $t \geq t_0$, provided that

$$V(t_0, x_0) \leq u_0 \quad \text{w.p. 1}. \qquad (4.11.2)$$

Let $0 < \varepsilon < \rho$ and $t_0 \in R_+$ be given. Since the solution $u \equiv 0$ of (4.8.19) is equistable with probability one, given $b(\varepsilon) > 0$, $t_0 \in J$, there exists a positive function $\delta_1 = \delta_1(t_0, \varepsilon)$ that is continuous in $t_0$ for each $\varepsilon$ such that $\sum_{i=1}^{N} u_{i0} \leq \delta_1$ implies

$$\sum_{i=1}^{N} u_i(t, t_0, u_0) < b(\varepsilon), \qquad t \geq t_0, \quad \text{w.p. 1}, \qquad (4.11.3)$$

where $u(t, t_0, u_0)$ is any solution process of (4.8.19). We choose $u_0$ so that

$$V(t_0, x_0) \leq u_0 \qquad \text{and} \qquad \sum_{i=1}^{N} u_{i0} = a(t_0, \|x_0\|) \quad \text{w.p. 1}. \qquad (4.11.4)$$

Since $a(t, \cdot) \in \mathcal{K}$, we can find a $\delta = \delta(t_0, \varepsilon)$ that is continuous in $t_0$ for each $\delta$ and satisfies the inequalities

$$\|x_0\| \leq \delta \qquad \text{and} \qquad a(t_0, \|x_0\|) \leq \delta_1 \quad \text{w.p. 1} \qquad (4.11.5)$$

simultaneously. We claim that $(\text{SS}_1)$ holds with this $\delta$. Suppose that this is not true. Then there would exist a solution process $x(t) = x(t, t_0, x_0)$ with $\|x_0\| \leq \delta$ w.p. 1 and a $t_1 > t_0$ such that

$$\|x(t_0)\| = \varepsilon, \qquad \|x(t)\| \leq \varepsilon, \qquad t \in [t_0, t_1], \quad \text{w.p. 1}, \qquad (4.11.6)$$

which implies $x(t) \in \{x \in R^n : \|x\| \leq \varepsilon\} \subset S(\rho)$ because of the choice of $\varepsilon$. From (4.11.1), (4.11.3), (4.11.4), (4.11.6), and hypothesis (iii), we are led to the contradiction

$$b(\varepsilon) = b(\|x(t_1)\|) \leq \sum_{i=1}^{N} V_i(t_1, x(t_1)) \leq \sum_{i=1}^{N} r_i(t_1, t_0, u_0) < b(\varepsilon) \quad \text{w.p. 1},$$

proving $(\text{SS}_1)$. The proof of the theorem is complete.

**Example 4.11.1.**  Consider the linear system of stochastic differential equations

$$dx = A(t)x\,dt + B(t)x\,dz, \qquad (4.11.7)$$

where $z(t)$ is a scalar Wiener process and $A$ and $B$ are $n \times n$ continuous matrix functions defined on $R_+$ into $R^{n^2}$. Take $V(x) = (x^{\mathrm{T}}Px)^{1/2}$, where $P$ is a positive definite matrix and $b(t)$ is the eigenvalue of the matrix $\frac{1}{2}P^{-1}[B^{\mathrm{T}}(t)P + PB(t)]$ of multiplicity $n$. Let $a(t)$ be the largest eigenvalue of $\frac{1}{2}P^{-1}[A^{\mathrm{T}}(t)P + PA(t) + B^{\mathrm{T}}(t)PB(t)]$. Assume that

$$P\left\{\limsup_{t \to \infty} \frac{1}{t - t_0}\left[\int_{t_0}^t (a(s) - b^2(s))\,ds + \int_{t_0}^t b(s)\,dz(s)\right] \le -\alpha\right\} = 1$$

for $\alpha > 0$. Then it is evident that

$$dV(x) = \alpha(t,x)\,dt + b(t)V(x)\,dz(t),$$

where

$$\alpha(t,x) \le \left[a(t) - \tfrac{1}{2}b^2(t)\right]V(x)$$

and

$$\alpha(t,x) = \tfrac{1}{2}(x^{\mathrm{T}}Px)^{-1/2}\left[x^{\mathrm{T}}(A^{\mathrm{T}}P + PA + B^{\mathrm{T}}PB)x\right]$$
$$- \tfrac{1}{8}(x^{\mathrm{T}}Px)^{-3/2}\left[x^{\mathrm{T}}(B^{\mathrm{T}}P + PB)x\right]^2.$$

Clearly, the trivial solution of $du = (a(t) - \tfrac{1}{2}b^2(t))u\,dt + b(t)u\,dz(t)$ is uniformly asymptotically stable with probability one. Consequently, the trivial solution of (4.11.7) is uniformly asymptotically stable with probability one.

Based on the proof of Theorem 4.11.1, it is not difficult to prove other a.s. stability properties. We leave the formulation of such results to the reader.

We further note that Lyapunov-like functions and stochastic differential inequalities of Itô-type can also be used to derive the stability results in probability as well as in the $p$th mean. In order to avoid monotony, we do not undertake the formulation of such results. However, we give a sufficient condition that illustrates the asymptotic stability in probability of the trivial solution of (4.11.7).

We assume that $a(t)$, $b(t)$, and $z(t)$ satisfy the relation

$$P\left\{\limsup_{b \to \infty} \frac{1}{t - t_0}\left[\int_{t_0}^t (a(s) - b^2(s))\,ds + \int_{t_0}^t b(s)\,dz(s)\right] \le 0\right\} = 1,$$

and that the trivial solution of $du = (a(t) - \tfrac{1}{2}b^2(t))u\,du + b(t)u\,dz(t)$ is uniformly asymptotically stable in probability. Consequently, the trivial solution of (4.11.7) is uniformly asymptotically stable in probability.

Finally, we shall give a simple result that will show the usefulness of the method of variation of parameters in the stability study of (4.8.1).

**Theorem 4.11.2.**  Assume that

(i)   $\sigma(t, x)$ in (4.8.1) is an $n \times 1$ matrix function such that $\sigma(t, x) \equiv B(t)x$ and $z$ is a scalar normalized Wiener process, where $B \in C[R_+, R^{n^2}]$ is an $n \times n$ matrix function;

(ii)   $f(t, 0) \equiv 0$ for $t \in R_+$;

(iii)   $\|f(t, x)\| = o(\|x\|)$ as $x \to 0$ uniformly in $t \in R_+$;

(iv)

$$P\left\{\omega : \limsup_{t \to \infty} \left[ \frac{1}{t - t_0} \left( \int_{t_0}^{t} b(s)\, ds + \left\| \int_{t_0}^{t} B(u)\, dz(u) \right\| \right) \right] \leq -a \right\} = 1$$

for some positive real number $a = a(t_0)$.

Then the trivial solution of (4.8.1) is asymptotically stable with probability one.

*Proof.*   By following the proof of Theorem 4.10.2, we arrive at (4.10.9). From hypothesis (iv) and following the argument (used in the proof of Theorem 2.12.3), the proof of the theorem can be completed very easily. We therefore leave the details to the reader.

**Problem 4.11.1.**   Assume that hypotheses (i)–(iii) of Theorem 4.11.2 hold. Further assume that $\beta(t, t_0)$ in Theorem 4.9.5 satisfies

$$P\left\{\omega : \limsup_{t \to \infty} \beta(t, t_0) \leq -\alpha \right\} = 1$$

for some positive number $a = a(t_0)$. Then the trivial solution of (4.8.1) is asymptotically stable with probability one.

## Notes

The Itô calculus presented in Section 4.1 follows the well-known texts on the theory of random processes, namely, Doob [14], Gikhman and Skorokhod [19], and Wong [87]. See also Itô [28–33], Friedman [18], and McShane [64]. Theorem 4.2.1 is based on the work on Itô and Nisio [34]. See also Gikhman and Skorokhod [20]. Theorem 4.2.2 is new. Theorem 4.2.3 is adapted from Arnold [1]. See also Bharucha-Reid [5], Gikhman and Skorokhod [20], Girsanov [21], Goldstein [22], Hardiman and Tsokos [23], Morozan [68], Ruymgaart and Soong [76], and Wong [87]. Theorem 4.4.2 is new. Theorems 4.3.1 and 4.3.2 are new and are direct extensions of the corresponding deterministic results given in Lakshmikantham and Leela [59]. Theorem 4.3.3 is based on the work of Gikhman and Skorokhod [20] and the deterministic results of Lakshmikantham and Leela [59]. Theorem 4.4.1 is taken from the work of Kulkarni and Ladde [47]. Theorem

4.4.2 is new and is an extension of the deterministic result of Lord and Mitchell [62]. The results of Sections 4.5 and 4.6 are adapted from Ladde and Lakshmikantham [56]. See also Ikeda and Watanabe [26], Gihman and Skorokhod [20], and Yamada [90]. The results of Section 4.7 are due to Ladde and Lakshmikantham [56]. Theorems 4.8.1, 4.8.2, and 4.8.3 are due to Ladde [50]. See also Ladde, Lakshmikantham, and Liu [54], Bucy [6], Cumming [13], and Kushner [48]. Theorem 4.8.4 is adapted from Ladde and Lakshmikantham [57]. Theorems 4.9.1, 4.9.2, and 4.9.3 are based on the work of Ladde, Lakshmikantham, and Liu [55]. This section illustrates the significance of the comparison result and related earlier stability results. Example 4.9.1, which is adapted from Ladde [50], demonstrates the usefulness of a vector Lyapunov function over a single Lyapunov function. For further related results, see Kleinman [42], Kozin [45], Nevel'son [70, 71], Nevel'son and Khas'minskii [73], and Wonham [88]. Theorem 4.10.1 is based on the result of Ladde. Lakshmikantham, and Liu [55]. See also Kushner [48], Ladde and Leela [58], McLane [63], Morazon [68], Miyahara [65], Nevel'son [71], Nevel'son and Khas'minskii [72, 73], and Zakai [91]. Theorem 4.11.1 and Example 4.11.1 are adapted from Ladde and Lakshmikantham [57]. The stability results in Sections 4.9, 4.10, and 4.11 that are based on the method of variation of parameters are developed on the basis of deterministic results in Lakshmikantham and Leela [59]. Problems connected with stochastic control theory, see Äström [2], Fleming and Rishel [17], and Wonham [89]. Furthermore, see also an introductory study of stochastic differential equations and their applications due to Srinivasan and Vasuderan [82].

# APPENDIX

## A.0. INTRODUCTION

The main objective of the following appendixes is to provide a brief survey of additional concepts and results from the theory of probabilitistic analysis which are used in the text or frequently used in the related literature. Section A.1 deals with moments of random functions and their applications in $L^2$-calculus. In Section A.2 a spectral representation of covariance and correlation functions is given. Section A.3 deals with some additional properties of Gaussian processes. Brownian motion and martingales are discussed in Section A.4 and A.5, respectively. In Section A.6 strictly stationary and metrically transitive processes are presented. Section A.7 deals with Markov processes and Kolmogorov backward and forward equations. Furthermore, Kolmogorov equations for diffusion processes are also considered. The study of Kolmogorov equations (backward and forward) provides the method to compute the so-called transition function. The nature of this problem has been well-known and well-recognized in the literature. Finally, in Section A.8 the closed graph theorem is stated.

## A.1. MOMENTS OF RANDOM FUNCTIONS

Some of the most important properties of random functions are characterized by their moments. We shall list some concepts related to the different kinds of moments of a stochastic process.

**Definition A.1.1.** Let $x \in R[[a,b], R[\Omega, R^n]]$, and let $f(t,x)$ be a density function of $x(t)$ for $t \in [a,b]$.

(i) *The mean or first moment* of a random function $x(t)$ is defined by

$$m(t) = E[x(t)] = \int_{R^n} xf(t,x)\,dx. \qquad (A.1.1)$$

**180**

(ii)   *The variance or second central moment* of $x(t)$ is defined by

$$V(t) = E[(x(t) - m(t))(x(t) - m(t))^{\mathrm{T}}] = \int_{R^n} (x - m)(x - m)^{\mathrm{T}} f(t, x) \, dx. \quad (A.1.2)$$

We note that the mean $m(t)$ is an $n$-vector function and the variance $V(t)$ is an $n \times n$ matrix function.

**Definition A.1.2.**   Let $x, y \in R[[a, b], R[\Omega, R^n]]$, and let $g(t, x; s, y)$ be a joint density function of the random functions $x(t)$ and $y(s)$ for $t, s \in [a, b]$. Furthermore, let $f(t, x_1; s, x_2)$ be a joint density function of $x(t)$ and $x(s)$ for $t, s \in [a, b]$.

(i)   *The cross-correlation function of* $x(t)$ *and* $y(s)$ *is defined by*

$$\Gamma_{xy}(t, s) = E[x(t)(y(s))^{\mathrm{T}}] = \int_{R^n} \int_{R^n} xy^{\mathrm{T}} g(t, x; s, y) \, dx \, dy. \quad (A.1.3)$$

(ii)   *The autocorrelation function of* $x(t)$ *is defined by*

$$\Gamma(t, s) = E[x(t)(x(s))^{\mathrm{T}}] = \int_{R^n} \int_{R^n} x_1 x_2^{\mathrm{T}} f(t, x_1; s, x_2) \, dx \, dx. \quad (A.1.4)$$

(iii)   *The cross-covariance function of* $x(t)$ *and* $y(s)$ *is defined by*

$$\begin{aligned} C_{xy}(t, s) &= E[(x(t) - m_x(t))(y(s) - m_y(s))^{\mathrm{T}}] \\ &= \int_{R^n} \int_{R^n} (x - m_x)(y - m_y)^{\mathrm{T}} g(t, x; s, y) \, dx \, dy. \end{aligned} \quad (A.1.5)$$

(iv)   *The autocovariance function of* $x(t)$ *is defined by*

$$\begin{aligned} C(t, s) &= E[(x(t) - m(t))(x(s) - m(s))^{\mathrm{T}}] \\ &= \int_{R^n} \int_{R^n} (x - m(t))(x - m(s))^{\mathrm{T}} f(t, x_1; s, x_2) \, dx_1 \, dx_2. \end{aligned} \quad (A.1.6)$$

(v)   *The correlation-coefficient function is defined by*

$$\rho(t, s) = C(t, s) / V(t) V(s). \quad (A.1.7)$$

**Remark A.1.1.**   We note that $\Gamma_{xy}(t, s)$, $\Gamma(t, s)$, $C_{xy}(t, s)$, and $C(t, s)$ are $n \times n$ matrix functions defined on $[a, b] \times [a, b]$ into $R^{n^2}$. Furthermore, $\Gamma(t, s) = C(t, s)$ if and only if $m(t) \equiv 0$ on $[a, b]$, and $\Gamma_{xy}(t, s) = C_{xy}(t, s)$ if and only if $m_x(t) \equiv m_y(t) \equiv 0$ on $[a, b]$.

The functions $\Gamma_{xy}$, $\Gamma$, $C_{xy}$, and $C$ in Definition A.1.2 satisfy the following properties:

$$H(t, s) = H(s, t) \qquad \text{for all} \quad t, s \in [a, b], \quad (A.1.8)$$

$$H^2(t, s) \leq H(t, t) H(s, s) \qquad \text{for all} \quad t, s \in [a, b], \quad (A.1.9)$$

where $H(t, s)$ stands for any one of the functions, $\Gamma_{xy}(t, s)$, $\Gamma(t, s)$, $C_{xy}(t, s)$, and $C(t, s)$.

The functions $\Gamma(t,s)$ and $C(t,s)$ satisfy the following property. The matrix functions $\Gamma(t,s)$ and $C(t,s)$ are nonnegative definite functions on $[a,b] \times [a,b]$, that is, for every $n$, $t_1, t_2, \ldots, t_n \in [a,b]$, and for an arbitrary deterministic function $g(t)$ defined on $[a,b]$,

$$\sum_{j,k=1}^{n} \Gamma(t_j, t_k)g(t_j)g(t_k) \geq 0, \tag{A.1.10}$$

and an inequality similar to (A.1.10) holds for $C(t,s)$.

**Example A.1.1.** Consider a random function defined by

$$x(t) = A\sin(bt + \theta),$$

where $A$ and $b$ are real constants and $\theta \in R[\Omega, R_+]$, moreover, it is uniformly distributed in the interval $[0, 2\pi]$. We obtain easily

$$E[x(t)] = E[A\sin(bt + \theta)] = \frac{A}{2\pi}\int_0^{2\pi}\sin(bt + s)\,ds = 0,$$

and

$$\Gamma(t,s) = A^2 E[\sin(bt + \theta)\sin(bs + \theta)] = \frac{A^2}{2}\cos(b(t-s)). \tag{A.1.11}$$

In the following we shall list certain properties of a random function $x \in R[[a,b], L^2[\Omega, R^n]]$ in terms of its autocorrelation or covariance functions.

**Theorem A.1.1.** Let $x$, $y \in R[[a,b], L^2[\Omega, R^n]]$. For $t_0, s_0 \in [a,b]$, let us assume $t \to t_0$ and $s \to s_0$ for $t, s \in [a,b]$, such that $x(t) \xrightarrow{\text{m.s.}} x(t_0)$ and $y(s) \xrightarrow{\text{m.s.}} y(s_0)$ as $t \to t_0$ and $s \to t_0$. Then

$$\lim_{\substack{t \to t_0 \\ s \to t_0}} E[x(t)(y(s))^T] = E[x(t_0)(y(s_0))^T].$$

**Theorem A.1.2.** Let $x_n \in R[[a,b], L^2[\Omega, R^n]]$ be a sequence of random functions. Then $x_n$ converges to $x \in R[[a,b], L^2[\Omega, R^n]]$ on $[a,b]$ if and only if the functions $E[x_n(t)(x_m(t))^T]$ converge to a finite function on $[a,b]$ for all $n$ and $m$ as $n, m \to n_0$, that is,

$$\Gamma_{x_n x_m}(t,s) \to \Gamma_{xx}(t,s) \qquad \text{as} \quad n \to \infty$$

on $[a,b] \times [a,b]$.

**Theorem A.1.3.** Let $x \in R[[a,b], L^2[\Omega, R^n]]$. Then $x \in C[[a,b], L^2[\Omega, R^n]]$ if and only if $\Gamma(t,s)$ is continuous at $(t,t)$.

**Theorem A.1.4.** If an autocorrelation function $\Gamma(t,s)$ of $x \in R[[a,b],$ $R[\Omega, R^n]]$ on $[a,b] \times [a,b]$ is continuous at every $(t,t) \in [a,b) \times [a,b]$, then it is continuous on $[a,b] \times [a,b]$.

**Theorem A.1.5.** Let $y(t)$ be a random function defined by

$$y(t) = \int_a^b G(t,s)x(s)\,ds, \tag{A.1.12}$$

where $G(t,\cdot) \in C[[a,b], R^{mn}]$, $x \in R[[a,b], L^2[\Omega, R^n]]$, and the integral is a Riemann integral in the mean square sense. $y(t)$ in (A.1.12) exists if and only if the ordinary double Riemann integral

$$\int_a^b \int_a^b G(t,u)\Gamma(u,s)(G(t,s))^{\mathrm{T}}\,du\,ds \tag{A.1.13}$$

exists and is finite.

This theorem suggests the definition of an autocorrelation function of $y(t)$ in (A.1.12).

**Definition A.1.3.** The autocorrelation function of $y(t)$ in (A.1.12) is given by

$$\Gamma_{yy}(t,s) = \int_a^b \int_a^b G(t,u)\Gamma(u,v)(G(s,v))^{\mathrm{T}}\,du\,dv. \tag{A.1.14}$$

Let us conclude this section by noting the following.

**Example A.1.2.** Let $y(t)$ be a random function defined by

$$y(t) = \int_a^t G(t,s)x(s)\,ds, \tag{A.1.15}$$

where $x \in R[[a,b], L^2[\Omega, R^n]]$ and $G(t,s) \in C[[a,b] \times [a,b], R^{mn}]$. By the application of Theorem A.1.5, one can conclude that $y(t)$ in (A.1.15) exists if and only if

$$\int_a^t \int_a^t G(t,u)\Gamma(u,s)(G(t,s))^{\mathrm{T}}\,du\,ds. \tag{A.1.16}$$

Therefore, the autocorrelation function of $y(t)$ in (A.1.15) is given by

$$\Gamma_{yy}(t,s) = \int_a^t \int_a^s G(t,u)\Gamma(u,v)(G(s,v))^{\mathrm{T}}\,du\,dv. \tag{A.1.17}$$

## A.2. SPECTRAL REPRESENTATIONS OF COVARIANCE AND CORRELATION FUNCTIONS

Here we discuss spectral representations of the autocovariance and autocorrelation functions of a stationary process.

**Definition A.2.1.** A random function $x \in R[[a,b], R[\Omega, R^n]]$ is said to be a broad- or wide-sense stationary process if $x \in R[[a,b], L^2[\Omega, R^n]]$, $E[x(t)] = m = \text{constant}$, and $E[x(t)(x(s))^T] = \Gamma(t,s)$ for all $t, s \in [a,b]$.

**Remark A.2.1.** We remark that for a stationary process $x(t)$ in the broad sense, the variance $V(t)$ of $x(t)$ is independent of $t$, that is, $V(t) = \Gamma(0)$ on $[a,b]$. It is obvious that a strictly stationary process that belongs to $R[[a,b], L^2[\Omega, R^n]]$ is also wide-sense stationary, but the converse is not true, for example, the Gaussian process.

We shall present additional properties of the autocorrelation and autocovariance functions of a wide-sense stationary process.

Let $x, y \in R[(-\infty, \infty), R[\Omega, R^n]]$ be a wide-sense stationary process, and let $\Gamma(t) = E[x(s)(x(s+t))^T]$ and $C(t) = E[(x(s) - m)(x(t+s) - m)^T]$ be the autocorrelation and autocovariance functions of $x(t)$, respectively. From (A.1.8)–(A.1.10) and the definitions of $\Gamma(t)$ and $C(t)$, it is obvious that

$$\Gamma(0) \geq 0 \quad \text{and} \quad C(0) \geq 0, \tag{A.2.1}$$

$\Gamma(t)$ and $C(t)$ are even functions on $(-\infty, \infty)$, that is,

$$\Gamma(t) = \Gamma(-t) \quad \text{and} \quad C(t) = C(-t) \quad \text{for} \quad t \in (-\infty, \infty), \tag{A.2.2}$$

and $\Gamma(t)$ and $C(t)$ are bounded functions on $(-\infty, \infty)$. Moreover,

$$|\Gamma(t)| \leq \Gamma(0) = E[x(s)(x(s))^T], \tag{A.2.3}$$

$$|C(t)| \leq C(0) = E[(x(s) - m)(x(s) - m)^T], \tag{A.2.4}$$

and they are nonnegative definite functions on $R$. Furthermore, if $\Gamma(t)$ and $C(t)$ are continuous at $t = 0$, then they are uniformly continuous on $(-\infty, -\infty)$.

In the following we state representation theorems for autocorrelation and autocovariance functions. To avoid monotony, we state only a representation theorem for $C(t)$.

**Theorem A.2.1** [19]. For an $n \times n$ matrix function to be the autocovariance function of $x \in R[R, R[\Omega, R^n]]$, which is stationary in the broad sense and satisfies the condition

$$\lim_{h \to 0} E[\|x(t+h) - x(t)\|^2] = 0, \tag{A.2.5}$$

it is necessary and sufficient that it have a representation of the form

$$C(t) = \int_{-\infty}^{\infty} \exp[iut] \, dF(u), \tag{A.2.6}$$

where $i = \sqrt{-1}$, $t \in R$, and $F(u) = (F_{ij}(u))$ is an $n \times n$ matrix function satisfying the following conditions: (a) for any $u_1, u \in R$ and $u_1 < u$, the matrix

$\Delta F(u) = F(u) - F(u_1)$ is nonnegative definite, and (b) $\mathrm{tr}(F(+\infty) - F(-\infty))$ is finite.

**Remark A.2.2.**   We remark that the positive definiteness of the matrix $\Delta F(u)$ implies that $|\Delta F_{ij}(u)|^2 \le \Delta F_{ii}(u) \Delta F_{jj}(u)$, so that

$$\sum_{k=1}^{m} |\Delta F_{ij}(u_k)| \le \left[ \sum_{k=1}^{m} \Delta F_{ii}(u_k) \right]^{1/2} \left[ \sum_{k=1}^{m} \Delta F_{jj}(u_k) \right]^{1/2}, \qquad (A.2.7)$$

where $\Delta F(u_k) = F(u_k) - F(u_{k-1})$ for $k = 1, 2, \ldots, m$ and $u_0 < u_1 < \cdots < u_m$. Furthermore, condition (a) implies that the diagonal elements of the matrix $F(u)$ are nondecreasing functions of $u$. The condition (b) is equivalent to the requirement that each diagonal element $F_{jj}(u)$ of the matrix $F(u)$ be of bounded variation on $R$. This, together with (A.2.7), implies that the off-diagonal elements $F_{ij}(u), i \ne j$, of the matrix $F(u)$ are also functions of bounded variation.

**Remark A.2.3.**   Relation (A.2.5) is equivalent to the continuity of the function $C(t)$ at $t = 0$.

**Remark A.2.4.**   If the function $F(u)$ in (A.2.6) is absolutely continuous on $R$, then we have $(dF/du)(u) = S(u)$ a.e. on $R$. Hence relation (A.2.6) reduces to

$$C(t) = \int_{-\infty}^{\infty} \exp[iut] \, dF(u). \qquad (A.2.8)$$

We further note that the $C(t)$ in (A.2.8) is the inverse Fourier transform of $S(u)$.

**Definition A.2.2.**   The spectral function of a wide-sense stationary random function $x \in R[R, R[\Omega, R^n]]$ is defined by the representation

$$C(t) = \int_{-\infty}^{\infty} \exp[iut] \, dF(u).$$

**Remark A.2.5.**   If we assume that the autocovariance function is absolutely integrable on $R$; that is, if $\int_{-\infty}^{\infty} \|C(t)\| \, dt < \infty$, then from (A.2.8) we have

$$S(u) = \frac{1}{2\pi} \int_{-\infty}^{\infty} \exp[-iut] C(t) \, dt. \qquad (A.2.9)$$

**Definition A.2.3.**   The spectral density function of a wide-sense stationary process $x \in R[R, R[\Omega, R^n]]$ is defined by (A.2.9).

**Remark A.2.6.**   We remark that $S(u)$ is a real and nonnegative function. Furthermore, $C(t)$ is an even function, and therefore relations (A.2.8) and (A.2.9) can be written as

$$C(t) = 2 \int_{-\infty}^{\infty} \cos(ut) S(u) \, du \qquad (A.2.10)$$

and

$$S(u) = \frac{1}{\pi} \int_{-\infty}^{\infty} \cos(ut)C(t)\,dt, \qquad (A.2.11)$$

respectively.

We further note that for $u = 0$, Eq. (A.2.11) reduces to

$$S(0) = \frac{1}{\pi} \int_{-\infty}^{\infty} C(t)\,dt. \qquad (A.2.12)$$

### A.3. SOME PROPERTIES OF GAUSSIAN PROCESSES

In view of the central limit theorem, a Gaussian process can be expected to occur whenever a sum of a random sample $x_1, x_2, \ldots, x_m$ (where $x_1$, $x_2, \ldots, x_m$ denote mutually independent random variables in $R[\Omega, R^n]$ which have the same, possibly unknown, distribution) is drawn from a population with finite mean $\mu$ and finite standard deviation $\sigma$. Therefore, we can expect that Gaussian processes are models of many physical and biological phenomena. The Wiener process or the process of Brownian motion, for example, is a Gaussian process.

Gaussian processes have a number of important properties that have several mathematical advantages.

**Theorem A.3.1.** A Gaussian process is completely specified by its mean and covariance functions.

**Theorem A.3.2.** Linear transformations of Gaussian processes map Gaussian distributions into Gaussian distributions. (If $x(t)$ is a Gaussian process, then $Ax(t)$ is also a Gaussian process, where $A$ is a linear mapping and $x \in R[I, R[\Omega, R^n]]$.)

**Theorem A.3.3.** Let $x_n(t) \in R[I, R[\Omega, R^n]]$ be a sequence of $n$-dimensional Gaussian processes having Gaussian distributions with parameters $m_n(t)$ and $V_n(t)$. The sequence of distributions of the processes $x_n(t)$ converges in distribution to some limiting distribution if and only if $m_n(t) \to m(t)$ and $V_n(t) \to V(t)$ for each $t \in I$. Then the limiting distribution is also a Gaussian distribution with parameters $m(t)$ and $V(t)$.

**Problem A.3.1.** Let $G$ be an $n \times m$ deterministic matrix function that belongs to $M_2[a, b]$, and let $z(t)$ be a normalized Wiener process. For any sub-$\sigma$-algebra $\mathscr{F}_t \supset \mathscr{Z}_t$, show that the Itô-type indefinite integral

$$y(t) = \int_a^t G(s)\,dz(s)$$

is a normally distributed process. Moreover, the mean is 0 and variance is $\int_a^t \|G(s)\|^2 \, ds$.

**Theorem A.3.4.**   Let $x \in R[a, b], R[\Omega, R^n]]$ be a Gaussian process, and let $y(t)$, defined by

$$y(t) = \int_a^t G(t, s)x(s) \, ds, \tag{A.3.1}$$

belong to $R[[a, b], L^2[\Omega, R^n]]$, where $G(t, \cdot) \in C[[a, b], R^{mn}]$ for $t \in [a, b]$. Then $y(t), t \in [a, b]$, is a Gaussian process.

**Theorem A.3.5.**   Let $x \in R[[a, b], R[\Omega, R^n]]$ be a Gaussian process, and let $x'(t)$ be the $L^2$-derivative of $x(t), t \in [a, b]$. Then $x'(t), t \in [a, b]$, is a Gaussian process.

**Remark A.3.1.**   From the preceding properties, it is obvious that the Gaussian properties are preserved under $L^2$-integration and differentiation.

**Example A.3.1.**   Let $x \in R[[a, b], R[\Omega, R^n]]$ be a Gaussian process. Then Theorem A.3.4 shows that the process defined by (A.3.1) is a Gaussian process. Now by Example A.1.2, it is obvious that the autocorrelation function $\Gamma(t, s)$ of $y(t)$ in (A.3.1) is given by

$$\Gamma(t, s) = \int_a^t \int_a^s G(t, u)\Gamma(u, v)(G(s, v))^T \, du \, dv, \tag{A.3.2}$$

provided that

$$\int_a^t \int_a^t G(t, u)\Gamma(u, s)(G(t, s))^T \, du \, ds$$

exists. If we assume that $E[x(t)] = 0$ on $[a, b]$, then by Remark A.1.1, the autocorrelation function $\Gamma(t, s)$ in (A.3.2) of $y(t)$ in (A.3.1) is an autocovariance function of $y(t)$. Therefore, the Gaussian process in (A.3.1) has zero mean and variance $V(t)$, where

$$V(t) = \int_a^t \int_a^t G(t, u)C(u, s)(G(t, s))^T \, du \, ds. \tag{A.3.3}$$

If we further assume that the Gaussian process is stationary in the broad or wide sense, then $\Gamma(t, s)$ in (A.3.2) reduces to $\Gamma(t - s)$, and thus the variance $V(t)$ in (A.3.3) of $y(t)$ in (A.3.1) becomes

$$V(t) = \int_a^t \int_a^t G(t, u)C(u - s)(G(t, s))^T \, du \, ds. \tag{A.3.4}$$

**Example A.3.2.**   Let $x \in R[[a, b], R[\Omega, R]]$ be a Gaussian process with $E(x(t)] \equiv 0$ on $[a, b]$ with its autocovariance function $C(t, s) = C(t - s)$. Let $y(t)$ be defined by

$$y(t) = \int_a^t g(s)x(s) \, ds, \tag{A.3.5}$$

where $g \in C[[a, b], R]$. From Example A.3.1, we infer that $y(t)$ in (A.3.5) is a Gaussian process with $E[y(t)] = 0$ on $[a, b]$ and variance $V(t)$, where

$$V(t) = \int_a^t \int_a^t g(u)C(u - s)g(s) \, du \, ds. \tag{A.3.6}$$

It is obvious that for $p \geq 1$,

$$E[\exp[py(t)]] = \frac{1}{\sqrt{2\pi V(t)}} \int_{-\infty}^{\infty} \exp\left[py - \frac{y^2}{2V(t)}\right] dy$$

$$= \exp[\tfrac{1}{2}p^2 V(t)]$$

$$= \exp\left[\tfrac{1}{2}p^2 \int_a^t \int_a^t g(u)C(u - s)g(s) \, du \, ds\right]. \tag{A.3.7}$$

The following result gives sufficient conditions for a process to be Gaussian.

**Theorem A.3.6.** Let $x \in C[[a, b], R[\Omega, R^n]]$ be a process with independent increments. Then $x(t)$ is a Gaussian process.

### A.4. BROWNIAN MOTION

A Wiener process or a process of Brownian motion $x \in R[I, R[\Omega, R^n]]$ is defined as a Gaussian process with independent increments. We note that for a process $x(t)$ with independent increments, the matrix function $D(s, t)$ defined by

$$D(s, t) = E[(x(t) - x(s) - m(t) + m(s))(x(t) - x(s) - m(t) + m(s))^{\mathrm{T}}]$$

satisfies the relation

$$D(s, t) = D(s, u) + D(u, t) \qquad \text{for} \quad s < u < t, \tag{A.4.1}$$

where $E[x(t)] = m(t)$.

If $x(t)$ is a process with stationary increments, then $D(s, t) = D(t - s)$, and (A.4.1) can be written as

$$D(s_1 + s_2) = D(s_1) + D(s_2), \qquad s_1, s_2 > 0. \tag{A.4.2}$$

If $D(t)$ is continuous, then it is obvious that the solutions of equation (A.4.2) are of the form

$$D(s) = Ds, \tag{A.4.3}$$

where $D$ is a nonnegative definite matrix. Formula (A.4.3) completely describes the structural function of a process of Brownian motion with stationary increments. Thus the process $x \in R[I, R[\Omega, R^n]]$ is defined as a homogeneous Gaussian process with independent increments such that $E[x(t)] = 0$ and $a^{\mathrm{T}}D(t)a = \sigma^2 t \|a\|^2$ for $a \in R^n$.

Let us now discuss a physical application of a Brownian motion process. Consider a sufficiently small particle (a molecule of gas or liquid) suspended in a liquid, and consider its motion under the influence of collisions with the molecules of the liquid, which are in chaotic thermal motion. This phenomenon in physics is known as "Brownian motion."

For the probabilistic analysis of this phenomenon, it is natural to assume that the velocities of the molecules with which the particle collides are random. If the liquid is homogeneous, then one can assume that the distribution of the velocities of the molecules are independent of the position of the molecules. It is obvious that the distribution may depend on the temperature, which is uniform everywhere. Let $x(t) = (x_1(t), x_2(t), x_3(t))$ be a position of a particle at a time $t$. If we further assume that the velocities of the different molecules are independent of each other and if we neglect the mass of the particle, then the displacement $x(t + h) - x(t)$ of the particle from a position $x(t)$ to $x(t + h)$ for any $t \geq 0$ and $h > 0$ will be independent of the position of the particle and its previous motion. This means that the position of the particle $x(t)$ is a process with independent increments. If the physical state of the liquid is independent of time, then it is easy to see that the process $x(t)$ is continuous and homogeneous relative to $t \geq 0$. By Theorem A.3.6, it is obvious that the process is a Gaussian process.

Now we shall assume that $x(0) = 0$. Suppose that $E[x(t)] = tm$ and $a^T D(t)a = t(a^t Aa)$, where $m \in R^3$ and $A$ is a $3 \times 3$ symmetric matrix. In the case of a homogeneous liquid in which there are no currents, the process must be isotropic (since the distribution of the projections of the velocities of the molecules of the liquid in an arbitrary direction is independent of that direction), that is, $(m, a)$ and $(Aa, a)$ are independent of $a$ for $\|a\| = 1$. This is possible when $(m, a) = 0$ and $(Aa, a) = \sigma^2 \|a\|^2$. Thus from the most general considerations of the physical phenomenon of Brownian motion, we have arrived at the definition of a Brownian motion process.

### A.5.  MARTINGALES

Let $x \in [R_+, R[\Omega, R^n]]$ such that $E[\|x(t)\|] < \infty$, and let $\{\mathscr{F}_t : t \in R_+\}$ be an increasing family of sub-$\sigma$-algebras of $\mathscr{F}$ such that $x(t)$ is $\mathscr{F}_t$-measurable and -integrable for all $t \in R_+$. Then $\{x(t), \mathscr{F}_t : t \in R_+\}$ is said to be martingale if $t > s$ implies

$$E[x(t)|\mathscr{F}_s] = x(s) \quad \text{w.p. 1} \qquad \text{for all} \quad t \in R_+.$$

The process is said to be a supermartingale (submartingale) if the above equality is replaced by

$$E[x(t)|\mathscr{F}_s] \leq x(s) \qquad (x(s) \leq E[x(t)|\mathscr{F}_s]).$$

For any Brownian motion $\{x(t): t \in [0, \infty)\}$, if we take $\mathscr{F}_t$ to be the smallest $\sigma$-algebra with respect to which $\{x(s); s \leq t\}$ are all measurable, then $\{x(t), \mathscr{F}_t : t \in [0, \infty)\}$ is a martingale. In fact,

$$
\begin{aligned}
E[x(t)|\mathscr{F}_s] &= E[(x(t) - x(s)) + x(s)|\mathscr{F}_s] \\
&= E[(x(t) - x(s))|\mathscr{F}_s] + E[x(s)|\mathscr{F}_s] \\
&= x(s) \quad \text{w.p. 1.}
\end{aligned}
$$

Let us list some important properties of martingales.

(i)   If $\{x(t), \mathscr{F}_t : t \in R_+\}$ is a submartingale and if $\Phi$ is a real function of the real variable $r$, which is monotone nondecreasing and convex, with $E\{\Phi(\|x(t_0)\|)\} < \infty$ for some $t_0 \in J$, then $\{\Phi(\|x(t)\|), \mathscr{F}_t, t \in R_+, t \leq t_0\}$ is a submartingale.

(ii)   If $\{x(t), \mathscr{F}_t : t \in R_+\}$ is a martingale, then $\{\|x(t)\|, \mathscr{F}_t : t \in R_+, t \leq t_0\}$ is a submartingale.

(iii)   Let $\{x(t): t \in R_+\}$ be a separable martingale such that $E[\|x(t)\|^2] < \infty$ for $t \in R_+$. Then

$$
P\left(\sup_{t \in J} \|x(t)\| \geq \varepsilon\right) \leq E\|x(b)\|^2 / \varepsilon^2.
$$

(iv)   For every martingale $x(t) \in L^p(\Omega, R^n)$ for $p > 1$, we have

$$
E\left[\sup_{t \in J} \|x(t)\|^p\right] \leq (p/(p - 1))^p E[\|x(b)\|^p].
$$

**Remark A.5.1.**   If $\{x(t): t \in R_+\}$ is a separable process with independent increments such that $P(x(t) - x(s) > 0) = P(x(t) - x(s) < 0), s \leq t$, then we have

$$
P\left(\sup_{t \in J} \|x(t)\| \geq \varepsilon\right) \leq 2P(\|x(b)\| \geq \varepsilon).
$$

**Problem A.5.1.**   Let $x \in R[[a, b], R[\Omega, R^n]]$ be a Markov process with a specified transition probability distribution function. That is, we suppose that there is a given function $P(s, x, t, y)$ that defines a Baire function of $x$ for fixed $s, t, y$ and defines a distribution function in $y$ for fixed $x, s, t$. Then show that the stochastic process defined by

$$
z(t) = P(t, x(t), y, b) = P\{x(b, \omega) \leq y | x(s) \text{ for } a \leq s \leq t\}
$$

is a martingale.

The following result shows that the Itô indefinite integral in Section 4.1 is a martingale.

**Proposition A.5.1.**   Let $z(t)$ be a normalized Wiener process and let $G$ be an $n \times m$ random matrix function belonging to $M_2[a,b]$. Define a process $x(t) = \int_a^t G(s)\, dz(s)$. Then the process $x(t)$ is a martingale for all $t \in [a,b]$.

## A.6.   METRICALLY TRANSITIVE PROCESSES

In this section we briefly summarize translation groups of a stationary process and metrically transitive processes.

**Definition A.6.1.**   A family $\{T_t : t \in R\}$ of transformations defined on $\Omega$ into itself is called a translation group of measure-preserving one-to-one point transformations if each $T_t$ is a one-to-one measure-preserving point transformation and $T_{s+t} = T_s T_t$.

Note that if $x$ is a random variable and if $\{T_t : t \in R\}$ is a translation group of measure-preserving one-to-one point transformation groups, the stochastic process $\{T_t x : t \in R\}$ is strictly stationary.

**Remark A.6.1.**   Each translation group of measure-preserving one-to-one point transformations induces a translation group of measure-preserving set transformations (see Doob [14, Chap. 10]).

**Definition A.6.2.**   Let $\{T_t : t \in R\}$ be a translation group of measure-preserving one-to-one point transformations, and let $A \in \mathscr{F}$. Consider the set $\{(t,\omega) : \omega \in T_t A\}$. If this set is $(t,\omega)$-measurable for each $A \in \mathscr{F}$, the translation group is said to be measurable.

**Definition A.6.3.**   If $x$ is a random variable, then the process defined by

$$x(t,\omega) = (T_t x)(\omega) = x(T_{-t}\omega)$$

is measurable in the pair $(t,\omega)$ if the translation group is measurable.

**Definition A.6.4.**   $A \in \mathscr{F}$ is called an invariant set under a translation group of measure-preserving point transformations if the set differs from its image under $T_t$ by $\Lambda_t$ such that $P(\Lambda_t) = 0$.

Note that the invariant sets form a Borel field of $\omega$-sets.

**Definition A.6.5.**   A random variable $x$ is called invariant under a translation group of measure-preserving point transformations $T_t$ if for each $t \in R$, $x = T_t x$ w.p. 1.

**Definition A.6.6.**   A translation group of measure-preserving point transformations is called metrically transitive if the only invariant sets are those that have probability 0 or 1, i.e., if the only invariant random variables are those that are indentically constant w.p. 1.

**Remark A.6.2.**   Definitions A.6.1–A.6.6 can be formulated for $t \in [0, \infty]$, in which case the word "group" is replaced by "semigroup."

From Definitions A.6.1–A.6.6 and Remarks A.6.1 and A.6.2, one can formulate the corresponding definitions for a translation group or semi-group of measure-preserving one-to-one set transformations. We illustrate a definition corresponding to Definition A.6.1.

**Definition A.6.7.**   A family $\{T_t : t \in R\}$ of transformations defined on $\Omega$ into itself is called a translation group of measure-preserving set transforma-tions if $T_{s+t} = T_s T_t$ holds except for the modulo sets of probability zero, that is, if for every $A \in \mathscr{F}$ and every choice of $T_s T_t A$, and this choice is one of the images of $A$ under $T_{s+t}$.

Let $x \in R[R, R[\Omega, R^n]]$ be strictly stationary, that is,

$$P\{\omega : x(t_1, \omega) \in B_1, x(t_2, \omega) \in B_2, \ldots, x(t_n, \omega) \in B_m\}$$
$$= P\{\omega : x(t_1 + h, \omega) \in B_1, x(t_2 + h, \omega) \in B_2, \ldots, x(t_n + h, \omega) \in B_m\}$$

for all $t_1, t_2, \ldots, t_n, h \in R$, and $B_1, B_2, \ldots, B_m$ Borel sets in $R^n$. Let $\mathscr{F}_x$ be the minimal $\sigma$-algebra containing the cylinder sets $A$, where

$$A = \{\omega : x(t_1, \omega) \in B_1, x(t_2, \omega) \in B_2, \ldots, x(t_n, \omega) \in B_m\}. \qquad \text{(A.6.1)}$$

Let

$$\tilde{\Omega} = \{\tilde{\omega} : \tilde{\omega} = \phi(t), \phi : R \to R^n\}.$$

Let $\tilde{\mathscr{F}}$ be the $\sigma$-algebra generated by the sets

$$\tilde{A} = \{\tilde{\omega} : \phi(t_1) \in B_1, \phi(t_2) \in B_2, \ldots, \phi(t_n) \in B_m\}, \qquad \text{(A.6.2)}$$

and let

$$U : \Omega \to \tilde{\Omega} \qquad \text{be defined by} \quad U\omega = \tilde{\omega} = x(t, \omega). \qquad \text{(A.6.3)}$$

From (A.6.1) and (A.6.2), we note that the inverse image of $\tilde{A} \in \tilde{\mathscr{F}}$ under $U$ is equal to $A \in \mathscr{F}_x$. Hence

$$\mathscr{F}_x = \{U^{-1}(\tilde{A}) : \tilde{A} \in \tilde{\Omega}\}.$$

For every $t \in R$, define a transformation

$$\tilde{T}_t : \tilde{\Omega} \to \tilde{\Omega}, \qquad \tilde{T}_t(\tilde{\omega}) = \phi(t + s) \qquad \text{(A.6.4)}$$

whenever

$$\tilde{\omega} = \phi(s) \qquad \text{for} \quad s \in R.$$

The transformation $\tilde{T}_t$ is called the shift transformation. Furthermore, it is defined on $\tilde{\Omega}$ into itself.

For every $A \in \mathscr{F}_x$ and $t \in R$, we define a set transformation $T_t : \tilde{\mathscr{F}} \to \tilde{\mathscr{F}}$ as

$$T_t(A) = U^{-1}(\tilde{T}_t(\tilde{A})), \qquad \text{where} \quad A = U^{-1}(\tilde{A}). \tag{A.6.5}$$

We note that the set $T_t(A)$ is not uniquely determined by $A$; however, it may be proved that $T$ is single valued, with modulo sets of probability zero. In fact, if $A_1 = T_t(A)$ and $A_2 = T_t(A)$, then $A_1$ is equivalent to $A_2$, that is,

$$P(A_1 \Delta A_2) = 0, \qquad \text{where} \quad A_1 \Delta A_2 = (A_1 - A_2) \cup (A_2 - A_1). \tag{A.6.6}$$

In addition, the family of transformations $T_t$ possesses the following properties:

$$P(T_t A) = P(A) \qquad \text{for every} \quad A \in F_x,$$

$$T_t \left( \bigcup_{i=1}^{\infty} A_i \right) = \bigcup_{i=1}^{\infty} T_t(A_i), \qquad A_i \in \mathscr{F}_x, \qquad i = 1, 2, \ldots,$$

and

$$T_t(\Omega - A) = \Omega - T_t(A).$$

We note that the above equations are valid w.p. 1. This shows that the family of transformations $T_t$ is a family of measure-preserving set transformations. Thus for each strictly stationary process, there corresponds a unique translation group or semigroup of measure-preserving set transformations. These transformations are called shift transformations, and the group or semigroup is called the shift group or semigroup. Sets and random variables invariant relative to the shift group or semigroup are called invariant sets and random variables of the process, respectively.

**Definition A.6.8.** Let $x \in R[R, R[\Omega, R^n]]$ be a strictly stationary process. The process $x(t)$ is called metrically transitive if the shift group is metrically transitive.

In the following we present some results that establish the metric transitivity of a stochastic process.

**Theorem A.6.1.** A process is metrically transitive if and only if the corresponding coordinate space process for which the shift transformations become point transformations is metrically transitive.

**Theorem A.6.2.** If $g : R^n \to R^N$ is Borel-measurable and if $x \in M[R, R[\Omega, R^n]]$ is a strictly stationary and metrically transitive process, then the process $y(t) = g(x(t))$ is a measurable, strictly stationary, and metrically transitive process.

*Proof.* Since $g$ is Borel-measurable and $x(t, \omega)$ is product-measurable, it follows that $g(x(t))$ is a product-measurable process. We have

$$P\{\omega : y(t_1 + h, \omega) \in B_1, y(t_2 + h, \omega) \in B_2, \ldots, y(t_m + h, \omega) \in B_m\}$$
$$= P\{\omega : x(t_1 + h, \omega) \in g^{-1}(B_1),$$
$$x(t_2 + h, \omega) \in g^{-1}(B_2), \ldots, x(t_m + h, \omega) \in g^{-1}(B_m)\}$$
$$= P\{\omega : x(t_1, \omega) \in g^{-1}(B_1),$$
$$x(t_2, \omega) \in g^{-1}(B_2), \ldots, x(t_m, \omega) \in g^{-1}(B_m)\}$$
$$= P\{\omega : y(t_1, \omega) \in B_1, y(t_2, \omega) \in B_2, \ldots, y(t_m, \omega) \in B_m\}$$

for all $t_1, t_2, \ldots, t_m$, $h \in R$, and $B_1, B_2, \ldots, B_m$ Borel sets in $R^N$. We note that $g$ is a Borel-measurable function and that therefore $g^{-1}(B_1), g^{-1}(B_2), \ldots,$ $g^{-1}(B_m)$ are Borel sets in $R^n$. The above identity proves that the function $y(t) = g(x(t))$ is strictly stationary. Let $\mathscr{F}_y$ be the minimal $\sigma$-algebra containing the cylinder sets corresponding to $y(t)$, and let the transformations $U_2$ and $T_t^2$ given in (A.6.3) and (A.6.5) correspond to the process $y(t)$. Since any cylinder set relative to the process $y(t)$ is a cylinder set relative to the process $x(t)$, we have $\mathscr{F}_y \subseteq \mathscr{F}_x$. We shall prove that for every $A \in \mathscr{F}_y$, we have $T_t(A)$ equivalent to $T_t^2(A)$.

Let $\mathscr{F}^* = \{A \in \mathscr{F}_x : T_t(A) \sim T_t^2(A) \text{ for all } t \in R\}$. $\mathscr{F}^*$ contains the cylinder sets relative to the process $y(t)$. In fact, letting

$$A = \{\omega : y(t_1, \omega) \in B_1^N, \ldots, y(t_m, \omega) \in B_m^N\},$$

we have $A = U_2^{-1}(\tilde{A})$, where $\tilde{A}$ is as in (A.6.2). Therefore, $T_t^2(A) = U_2^{-1}(\tilde{T}_t \tilde{A})$. But

$$\tilde{T}_t(\tilde{A}) = \{\tilde{\omega} : \phi(t_1 + h) \in B_1, \phi(t_2 + h) \in B_2, \ldots, \phi(t_m + h) \in B_m\}$$

and

$$U_2^{-1}(\tilde{T}_t(\tilde{A})) = \{\omega : y(t_1 + h) \in B_1, y(t_2 + h) \in B_2, \ldots, y(t_m + h) \in B_m\}$$
$$= \{\omega : x(t_1 + h) \in g^{-1}(B_1),$$
$$x(t_2 + h) \in g^{-1}(B_2), \ldots, x(t_m + h) \in g^{-1}(B_m)\}$$
$$= U^{-1}(\tilde{A}_1),$$

where

$$\tilde{A}_1 = \{\tilde{\omega} : \phi(t_1 + h) \in g^{-1}(B_1),$$
$$\phi(t_2 + h) \in g^{-1}(B_2), \ldots, \phi(t_m + h) \in g^{-1}(B_m)\}.$$

Hence

$$U_2^{-1}(\tilde{T}_t(\tilde{A})) = U^{-1}(\tilde{A}_1) = U^{-1}(\tilde{T}_t(\tilde{A}_2)),$$

where

$$\tilde{A}_2 = \{\tilde{\omega}: \phi(t_1) \in g^{-1}(B_1),\ \phi(t_2) \in g^{-1}(B_2),\ \ldots,\ \phi(t_m) \in g^{-1}(B_m)\}.$$

On the other hand $A = U^{-1}(\tilde{A}_2)$, and hence $T_t(A) = U^{-1}(\tilde{T}_t(\tilde{A}_2))$. Consequently, by (A.6.6), $T_t(A)$ is equivalent to $T_t^2(A)$ and therefore contains the cylinder sets corresponding to the process $y(t)$. From the properties of $T_t$ in (A.6.5), it is obvious that $\mathscr{F}^*$ is a $\sigma$-algebra. Since $\mathscr{F}^*$ is a $\sigma$-algebra containing the cylinder sets corresponding to $y(t)$, it follows from the choice of $\mathscr{F}_y$ that $\mathscr{F}^* = \mathscr{F}_y$. Thus for every $A \in \mathscr{F}_y$, we have $T_t(A) \sim T_t^2(A)$ for all $t \in R$.

Let $A \in \mathscr{F}_y$ be an invariant set relative to $y(t)$. Then $T_t^2(A) \sim A$ for all $t \in R$. But $T_t(A) \sim T_t^2(A)$ implies that $T_t(A) \sim A$ for all $t \in R$. Hence $A$ is an invariant set relative to the process $x(t)$. Furthermore, the probability of $A$ is either zero or one in view of the fact that $x(t)$ is a metrically transitive process. By definition, this implies that $y(t)$ is a metrically transitive process.

## A.7.   MARKOV PROCESSES

Several phenomena that take place in physical and biological sciences are random processes. Many phenomena require the ability to compute the so-called transition function $P(s, x; t, B)$ expressing the probability that a system with initial state $x$ at time $s$ arrives in the set $B$ at time $t$. In 1931 Kolmogorov, under certain general conditions, showed that this probability can be calculated from certain differential equations.

Let us consider a Markov process $x \in R[[a, b],\ R[\Omega, R^n]]$ with fixed $\sigma$-algebra $\mathscr{F}$ and state space $(R^n, \mathscr{F}^n)$. Let $P(s, x; t, B)$ denote the transition function for $s < t$, where $s, t \in [a, b]$, $x \in R^n$, and $B \in \mathscr{F}^n$. For $s = t$, we have $P(s, x; x, B) = I_B(x)$, where $I_B(x)$ is a characteristic function of the set $B$.

The investigation of the behavior of the trajectories of a Markov process $x \in R[[a, b],\ R[\Omega, R^n]]$ requires a strengthening of the Markov process.

**Definition A.7.1.**   The process $x \in R[[a, b],\ R[\Omega, R^n]]$ is said to be a strong Markov process if it is measurable and if

(i)   $P(\cdot, \cdot; t, B)$ is an $(\mathscr{F}^1/[a, b] \times \mathscr{F}^n)$-measurable function for fixed $t \in [a, b]$ and $B \in \mathscr{F}^n$, where $\mathscr{F}^n/[a, b]$ is the $\sigma$-algebra of Borel sets in $[a, t]$;
(ii)   for an arbitrary Markov time $\tau$ of $x(t), t \in [a, b], P^\tau(x(t + \tau) \in B/\mathscr{F}_\tau) = P(\tau, x(\tau); \tau + t, B)$, $P^\tau$ a.s. for all $t \in [a, b]$, $B \in \mathscr{F}^n$.

**Definition A.7.2.**   For a Markov process $x \in R[[a, b],\ R[\Omega, R^n]]$, assume that

$$\lim_{h + k \to 0^+} \left[ \frac{P(t - h, x; t + k, B) - I_B(x)}{h + k} \right] \tag{A.7.1}$$

exists, and denote it by $Q(t, x, B)$. This function $Q(t, x, B)$ is called the transition intensity function corresponding to $P(s, x; t, B)$.

The transition intensity function $Q(t, x, B)$ satisfies the following properties:

(i)  $Q$ is completely additive with respect to $B \in \mathscr{F}^n$,
(ii)  $Q$ is $\mathscr{F}^n$-measurable with respect to $x \in R^n$, and

$$
\text{(iii)} \quad Q(t, x, B) \begin{cases} \leq 0 & \text{if} \quad x \in B, \\ = 0 & \text{if} \quad B = \emptyset, \\ \geq 0 & x \notin B. \end{cases}
$$

Let us represent $Q(t, x, B)$ in terms of two other functions that have an interesting probabilistic interpretation. Set

$$
q(t, x) = -Q(t, x, \{x\})
$$

and

$$
\sigma(t, x, B) = \begin{cases} I_B(x) & \text{if} \quad q(t, x) = 0, \\ \dfrac{Q(t, x, B - \{x\})}{q(t, x)} & \text{if} \quad q(t, x) \neq 0, \end{cases}
$$

where $x \in R^n$, $B \in \mathscr{F}^n$, and $t \in [a, b]$.

It is obvious that the properties of $Q(t, x, B)$ imply that (i) $q$ is nonnegative and $\mathscr{F}^n$-measurable with respect to $x \in R^n$ and (ii) $\sigma$ is nonnegative, $\mathscr{F}^n$-measurable relative to $x \in R^n$, and completely additive with respect to $B \in \mathscr{F}^n$ with $\sigma(t, x, R^n) \leq 1$. Furthermore,

$$
Q(t, x, B) = q(t, x)[-I_B(x) + \sigma(t, x, B)]. \tag{A.7.2}
$$

The probabilistic interpretations of $q$ and $\sigma$ are as follows:

$$
P(s, x; t, \{x\}) = 1 - (q(s, x) + o(1))(t - s); \tag{A.7.3}
$$

consequently, $q(s, x)(t - s) + o(t - s)$ is the probability of no longer being in $x$ at time $t$, given that the process was in $x$ at time $s$. Finally, if $q(s, x) \neq 0$, we may write

$$
\sigma(s, x, B) = \lim_{t \to s^+} \left[ \frac{P(s, x; t, B - \{x\})}{1 - P(s, x; t, \{x\})} \right]. \tag{A.7.4}
$$

Thus $\sigma(s, x, B)$ might be thought of as the probability of a jump from $x$ at time $s$ into the set $B - \{x\}$, given that a jump has in fact occurred.

**Theorem A.7.1**  (Kolmogorov's backward equation).  Assume that

$$
\lim_{h+k \to 0^+} \left[ \frac{P(t - h, x; t + k, B) - I_B(x)}{h + k} \right]
$$

exists for $t \in [a, b]$, $x \in R^n$, and $B \in \mathscr{F}^n$. Then the partial derivative $(\partial P/\partial s)(s, x; t, B)$ exists for almost all $s \in [a, b]$ (the set of expected values of $s$ may depend on $x \in R^n$, $t \in [a, b]$, and $B \in \mathscr{F}^n$) and satisfies the equation

$$\frac{\partial P}{\partial s}(s, x; t, B) = -\int_{R^n} Q(s, x, dy) P(s, y; t, B), \qquad (A.7.5)$$

and moreover,

$$\frac{\partial P}{\partial s}(s, x; t, B) = q(s, x) P(s, x; t, B) - \int_{R^n - \{x\}} Q(s, x, dy) P(s, y; t, B). \quad (A.7.6)$$

Equations (A.7.5) and (A.7.6) are called the Kolmogorov backward equations because they involve differentiation with respect to the earlier time $s$. The following result gives a necessary condition to obtain the Kolmogorov forward equation, that is, the equation that arises by differentiation with respect to the later time $t$.

**Theorem A.7.2** (Kolmogorov's forward equation). Assume that the hypotheses of Theorem A.7.1 hold. Further assume that the function $q(t, x) = -Q(t, x, \{x\})$ is bounded in $x \in R^n$ for each $t \in [a, b]$. Then the partial derivative $(\partial P/\partial t)(s, x; t, B)$ exists for almost all $t \in [a, b]$ (the set of expected values of $t$ may depend on $s \in [a, b]$, $x \in R^n$, and $B \in \mathscr{F}^n$) and satisfies the equation

$$\frac{\partial P}{\partial t}(s, x; t, B) = \int_{R^n} P(s, x; t, dy) Q(t, y, B). \qquad (A.7.7)$$

**Remark A.7.1.** To obtain Eqs. (A.7.5) and (A.7.7), stronger restrictions must be imposed on $P(s, x; t, B)$ and the limit (A.7.1) for each $s, t \in [a, b]$. Thus to deduce the backward equation it is enough to assume that $P(\cdot, x; t, B)$ is right-continuous uniformly with respect to $x \in R^n$ and that the limit in (A.7.1) is uniform with respect to $B \in \mathscr{F}^n$. Similarly, to derive the forward equation, it suffices to assume that $P(s, x; \cdot, B)$ is left-continuous on $[a, b]$ uniformly with respect to $B \in \mathscr{F}^n$ and that the limit in (1.7.1) is uniform with respect to $x \in R^n$.

**Example A.7.1.** Let $x \in R[[a, b], R[\Omega, R^n]]$ be a Markov process that takes a finite set of values, say $\{1, 2, \ldots, m\}$. The transition function

$$P(s, x; t, A) = P(s, i; t, \{j\})$$

will be determined by the transition matrix function

$$P(s, t) = (p_{ij}(s, t)), \qquad (A.7.8)$$

where

$$p_{ij}(s, t) = P(s, i; t, \{j\}). \qquad (A.7.9)$$

Clearly, we have $p_{ij}(s, t) \geq 0$, $\sum_{j=1}^{n} p_{ij}(s, t) = 1$.

The Chapman–Kolmogorov equation reduces to

$$p_{ij}(s, t) = \sum_{k=1}^{m} p_{ik}(s, u) p_{kj}(u, t)$$

for $a \le s \le u \le t \le b$ with $p_{ij}(s, s) = \delta_{ij}, s \in [a, b]$, where $\delta_{ij} = 0$ if $i \ne j$ and $\delta_{ij} = 1$ if $i = j$. Suppose that $p_{ij}(s, t) \in C[[a, b] \times [a, b], R]$ for all $i, j = 1, 2, \ldots, m$. Then it is clear that

$$\lim_{h+k \to 0^+} \left[ \frac{P(t - h, i; t + k, \{j\}) - I_{\{j\}}(i)}{h + k} \right] \tag{A.7.10}$$

exists and is equal to $q_{ij}(t)$. Moreover, $(\partial p/\partial s)ij$ and $(\partial p/\partial t)ij$ exist for all $i, j = 1, 2, \ldots, m$.

The matrix function $Q(s, t) = (q_{ij}(s, t))$ is said to be the transition intensity matrix function corresponding to $P(s, t)$. It is obvious that

$$q_{ij}(t) \begin{cases} \le 0 & \text{if} \quad i = j, \\ \ge 0 & \text{if} \quad i \ne j, \end{cases}$$

and

$$\sum_{j=1}^{m} q_{ij}(t) = 0 \qquad \text{for} \quad i \ge 1 \quad \text{and} \quad t \in [a, b].$$

Thus, assuming the continuity of the $p_{ij}(s, t)$ and the existence of $Q(t)$, we can write the backward and the forward equations (A.7.5) and (A.7.7), respectively, as follows:

$$\frac{\partial P}{\partial s}(s, t) = Q(s)P(s, t), \qquad P(t, t) = I, \tag{A.7.11}$$

$$\frac{\partial P}{\partial t}(s, t) = P(s, t)Q(t), \qquad P(s, s) = I. \tag{A.7.12}$$

**Remark A.7.2.** If a Markov jump process with discrete state is homogeneous, then the limit in (A.7.10) exists without any additional assumptions.

**Remark A.7.3.** Based on our discussion in Example A.7.1 and the representation of $Q(t, x, B)$ as a function of $q(t, x)$ and $\sigma(t, x, B)$, one can derive the backward and forward equations with regard to a Markov jump process with a denumerable number of values.

The objective of determining the transition function for a Markov function reduces to determining the answers to the following problems: (a) the existence of solutions of (A.7.5) and (A.7.7), (b) the uniqueness of solutions of (A.7.5) and (A.7.7), and (c) finding a solution or an estimate for solutions of

Eqs. (A.7.5) and (A.7.7). These problems (a)–(c) are equivalent to the problem of deterministic problems of ordinary integrodifferential equations.

The determination of a transition function of an a.s. continuous Markov process (diffusion process) $x \in R[[a, b], R[\Omega, R^n]]$ is equivalent to the problem of solving a certain kind of partial differential equation of parabolic type.

We give simple examples of diffusion processes:

(1)   Uniform motion with velocity $u$ is a diffusion process with $f \equiv u$ and $b \equiv 0$.

(2)   The Wiener process $w(t)$ is a diffusion process with drift vector $f \equiv 0$ and diffusion matrix $b \equiv I$.

To each diffusion process with coefficients $f$ and $b$ there is assigned the second order differential operator

$$L = f(t, x) \frac{\partial}{\partial x} + \frac{1}{2} \operatorname{tr} \left( b \frac{\partial^2}{\partial x \, \partial x} \right). \tag{A.7.13}$$

where tr stands for the trace of a matrix. This operator can be calculated as follows:

$$dg(s, x) = \lim_{t \to 0^+} \frac{1}{t} \left[ \int_{R^n} (g(s + t, y) - g(s, x)) P(s, x, t + s, dy) \right] \tag{A.7.14}$$

by means of a Taylor expansion of $g(s + t, y)$ about $(s, x)$ under the assumption that $g$ is defined and bounded on $I \times R^n$ and is twice continuously differentiable with respect to $x$ and once continuously differentiable with respect to $s$. Together with relations (4.1.27) and (4.1.28), the right-hand member of (A.7.14) gives the operator $(\partial/\partial s) + L$, where $d = (\partial/\partial s) + L$. For the time-independent function $g$ and the homogeneous case $d = L$.

We shall next state the following well-known results.

**Theorem A.7.3.**   Let $x(t)$, $t \in I$, be an $n$-dimensional diffusion process with continuous coefficients $f$ and $b$. The limit relations in (4.1.26)–(4.1.28) hold uniformly in $s \in I$. Let $g(x)$ denote a continuous bounded scalar function such that

$$u(s, x) = E[g(x(t)) | x(s) = x] = \int_{R^n} g(y) P(s, x, t, dy)$$

has bounded continuous first and second derivatives with respect to $x$. Then $u(s, x)$ is differentiable with respect to $s$ and satisfies Kolmogorov's backward equation

$$\frac{\partial u}{\partial s} + Lu = 0,$$

where $L$ is as defined in (A.7.13), with boundary condition

$$\lim_{s \to t} u(s, x) = g(x).$$

**Theorem A.7.4.** Let the assumptions of Theorem A.7.3 regarding $x(t)$ be satisfied. If $P(s, x; t, \cdot)$ has a density $p(s, x; t, y)$ that is continuous with respect to $s$ and if the derivatives $\partial p/\partial x$, $\partial^2 p/\partial x\, \partial x$ exist and are continuous with respect to $s$, then $p$ is called a fundamental solution of the backward equation

$$\frac{\partial p}{\partial s} + Lp = 0,$$

and it satisfies the boundary condition

$$\lim_{s \to t} p(s, x; t, y) = \delta(x - y),$$

where $\delta$ is Dirac's delta function.

**Theorem A.7.5.** Let $x(t)$ be an $n$-dimensional diffusion process for which the limit relations (4.1.26)–(4.1.28) hold uniformly in $s$ and $x$, and let it possess a transition density $p(s, x; t, y)$. Furthermore, assume that $\partial p/\partial t$, $(\partial/\partial y)(fp)$, $(\partial^2/\partial y\, \partial y)(bp)$ exist and are continuous functions. Then for fixed $s$ and $x$ such that $s \leq t$, this transition density $p(s, x; t, y)$ is a fundamental solution of the Kolmogorov forward equation or the Fokker–Plank equation

$$\frac{\partial p}{\partial t} + \sum_{i=1}^{n} \frac{\partial}{\partial y_i} (f_i(t, y)p) - \frac{1}{2} \sum_{i=1}^{n} \sum_{j=1}^{n} \frac{\partial^2}{\partial y_i\, \partial y_i} (b_{ij}(t, y)p) = 0 \quad \text{(A.7.15)}$$

**Example A.7.2.** For the Wiener process, the forward equation (A.7.15) for the homogeneous transition density

$$p(t, x, y) = (2\pi t)^{-1/2} \exp\left[ -(2t)^{-1} \| y - x \|^2 \right]$$

becomes

$$\frac{\partial p}{\partial t} = \frac{1}{2} \sum_{i=1}^{n} \frac{\partial^2 p}{\partial y_i^2},$$

which in this case is identical to the backward equation with $y$ replaced by $x$.

## A.8. CLOSED GRAPH THEOREM

Let $X$ be a normed linear space. A linear operator $T$ with domain $D(T) \subseteq X$ is said to be closed if whenever $x_n \to x$ as $n \to \infty$, $x_n \in D(A)$ and whenever $Ax_n \to y$ as $n \to \infty$, then $x \in D(A)$ and $Ax = y$.

**Theorem A.8.1** (Closed graph theorem). A closed linear operator mapping a Banach space into a Banach space is bounded (and thus continuous).

## Notes

Section A.1 is based on the material in the book of Soong [81]. Theorem A.2.1 is adapted from Gikhman and Skorokhod [19]. The results of Section A.3 are also taken from Soong [81]. For Theorem A.3.6, refer to Iosifescu and Tâutu [27] and Arnold [1]. Section A.4 is based on the discussion in the book of Gikhman and Skorokhod [19]. Section A.5 is taken from Doob [14]; see also Arnold [1]. The material in Section A.6 is formulated on the basis of the discussion in Doob [14]. Theorem A.6.2 is adapted from Morozan [66]. The results in Section A.7 are contained in Iosifescu and Tâutu [27] and Doob [14]. Theorem A.8.1 is based on the result in Dunford and Schwartz [15].

# REFERENCES

[1]   Arnold, L., "Stochastic Differential Equations: Theory and Applications." Wiley, New York, 1974.

[2]   Åström, K. J., "Introduction to Stochastic Control Theory." Academic Press, New York, 1970.

[3]   Bartlett, M. S., "Introduction to Stochastic Processes," 2nd ed. Cambridge University Press, London, 1966.

[4]   Bertram, J. E., and Sarachik, P. E., "Stability of circuits with randomly time-varying parameters. *IRE Trans. Speeial Suppl. PGIT-5*, (1959), 260–270.

[5]   Bharucha-Reid, A. T., "Random Integral Equations." Academic Press, New York, 1972.

[6]   Bucy, R. S., Stability and positive supermartingales. *J. Differential Equations* (1965), 151–155.

[7]   Bunke, H., Stabilität bei stochastischen Differentialgleichungssystemen. *Z. Angew. Math. Mech.* **43** (1963), 63–70.

[8]   Bunke, H., Über die fast sichere Stabilität linear stochastischer Systeme. *Z. Angew. Math. Mech.* **43** (1963), 533–535.

[9]   Bunke, H., Über das asymptotisch Verhalten von lösungen linearer stochastischer Differentialgleichungssysteme. *Z. Angew. Math. Mech.* **45** (1965), 1–9.

[10]  Bunke, H., On the stability of ordinary differential equations under persistent random disturbances. *Z. Angew. Math. Mech.* **51** (1971), 543–546.

[11]  Caughey, T. K., Comments on On the stability of random systems. *J. Acoust. Soc. Amer.* **32** (1960), 1356.

[12]  Caughey, T. K., and Gray, A. H., Jr., On the almost sure stability of linear dynamic systems with stochastic coefficients. *ASME J. Appl. Mech.* **32** (1965), 365–372.

[13]  Cumming, I. G., Derivation of the moments of a continuous stochastic system. *Internat. J. Control* **5** (1967), 85–90.

[14]  Doob, J. L., "Stochastic Processes." Wiley, New York, 1953.

[15]  Dunford, N., and Schwartz, J., "Linear Operators," Vol. I. Wiley (Interscience), New York, 1957.

[16]  Edsinger, R. W., Random Ordinary Differential Equations. Doctoral dissertation, University Of California, Berkeley, California, 1968.

[17]  Fleming, W. H., and Rishel, R. W., "Deterministic and Stochastic Optimal Control." Springer-Verlag, Berlin and New York, 1975.

[18]  Friedman, A., "Stochastic Differential Equations and Applications," Vols. I and II. Academic Press, New York, 1975.

[19] Gikhman, I. I., and Skorokhod, A. V., "Introduction to the Theory of Random Processes." Saunders, Philadelphia, Pennsylvania, 1969.

[20] Gikhman, I. I., and Skorokhod, A. V., "Stochastic Differential Equations." Springer-Verlag, Berlin and New York, 1972.

[21] Girsanov, I. V., An example of non-uniqueness of the stochastic equation of K. Itô. *Theor. Probability Appl.* **7** (1962), 325–331.

[22] Goldstein, J. A., An existence theorem for linear stochastic differential equations. *J. Differential Equations* **3** (1967), 78–87.

[23] Hardiman, S. T., and Tsokos, C. P., Existence theory for a stochastic differential equation. *Internat. J. Systems Sci.* **5** (1974), 615–621.

[24] Hartman, P., "Ordinary Differential Equations." Wiley, New York, 1964.

[25] Hille, E., and Phillips, R. S., "Functional Analysis and Semi-Groups." Amer. Math. Soc. Coll. Publ. Vol. 31. Amer. Math. Soc., Providence, Rhode Island, 1957.

[26] Ikeda, N., and Watanabe, S., A comparison theorem for solutions of stochastic differential equations and its applications." *Osaka J. Math.* **14** (1977), 619–633.

[27] Iosifescu, M., and Tăutu, P., "Stochastic Processes and Applications in Biology and Medicine," Vol. 1, Theory. Springer-Verlag, Berlin and New York, 1973.

[28] Itô, K., On stochastic processes. *Japan. J. Math.* **18** (1942), 261–524.

[29] Itô, K., Stochastic integral. *Proc. Imp. Acad. Tokyo* **20** (1944), 519–524.

[30] Itô, K., On stochastic integral equations. *Proc. Japan Acad.* **22** (1946), 32–35.

[31] Itô, K., Stochastic differential equations in a differentiable manifold. *Nagoya Math. J.* **1** (1950), 35–47.

[32] Itô, K., On a formula concerning stochastic differentials. *Nagoya Math. J.* **3** (1951), 55–65

[33] Itô, K., On Stochastic Differential Equations. *Mem. Amer. Math. Soc.*, **4**, (1951).

[34] Itô, K., and Nisio, M., On stationary solutions of a stochastic differential equation. *J. Math. Kyoto Univ.* **4** (1964), 1–75.

[35] Kats, I. Ia., On the stability of stochastic systems in the large. *Prikl. Mat. Meh.* **28** (1964), 366–372.

[36] Kats, I. Ia., and Krasovskii, N. N., On the stability of systems with random parameters' *Prikl. Mat. Meh.*, **24** (1960), 809–823.

[37] Khas'minskii, R. Z., On the stability of the trajectory of Markov processes. *Prikl. Mat. Meh.*, **26** (1962), 1025–1032.

[38] Khas'minskii, R. Z., On the dissipativity of random processes defined by differential equations. *Problemy Peredăchi Informatcii* **1** (1965), 84–104.

[39] Khas'minskii, R. Z., On the stability of nonlinear stochastic systems. *Prikl. Mat. Meh.* **30** (1966), 915–921.

[40] Khas'minskii, R. Z., Stability in the first approximation for stochastic systems. *Prikl. Mat. Meh.* **31** (1967), 1021–1027.

[41] Khas'minskii, R. Z., "Ustoychivost'sistem differentsial'nykh uravneniy pri sluchaynykh vozmusheniyakh (Stability of systems of differential equations in the presence of random disturbances)." Moscow, Nauka Press, (1969).

[42] Kleinman, D. L., On the Stability of Linear Stochastic Systems. *IEEE Trans. Automatic Control AC-14* (1969), 429–430.

[43] Kozin, F., On almost sure stability of linear systems with random parameters. *J. Math. Phys.* **43** (1963), 59–67.

[44] Kozin, F., On the relations between moment properties and almost sure Lyapunov stability for linear stochastic systems. *J. Math. Anal. Appl.* **10** (1965), 342–352.

[45] Kozin, F., On almost sure asymptotic sample properties of diffusion processes defined by stochastic differential equation, *J. Math. Kyoto Univ.* **4** (1965), 515–528.

[46] Kozin, F., A survey of stability of stochastic systems, *Automatica*-J. IFAC. **5** (1969), 95–112.

[47] Kulkarni, R. M., and Ladde, G. S., Stochastic Perturbations of Nonlinear Systems of Differential Equations. *J. Mathematical and Physical Sci.* **10** (1976), 33–45.

[48] Kushner, H. J., "Stochastic Stability and Control." Academic Press, New York, 1967.

[49] Ladas, G., and Lakshmikantham, V., "Differential Equations in Abstract Spaces." Academic Press, New York, 1972.

[50] Ladde, G. S., Systems of differential inequalities and stochastic differential equations II. *J. Mathematical Phys.* **16** (1975), 894–900.

[51] Ladde, G. S., Systems of differential inequalities and stochastic differential equations III. *J. Mathematical Phys.* **17** (1976), 2113–2120.

[52] Ladde, G. S., Logarithmic norm and stability of linear systems with random parameters. *Internat. J. Systems Sci.* **8** (1977), 1057–1066.

[53] Ladde, G. S., Stability technique and thought provocative dynamical systems II. In "Applied Nonlinear Analysis" (V. Lakshmikantham, ed.), pp. 215–218. Academic Press, New York, 1979.

[54] Ladde, G. S., Lakshmikantham, V. and Liu, P. T., Differential inequalities and Itô type stochastic differential equations. *Proc. Int. Conf. Nonlinear Differential and Functional Equations, Bruxelles et Louvain, Belgium, September 3–8, 1973* (P. Janssens, J. Mawhin, and N. Rouche, eds.), pp. 611–640. Herman, Paris, 1973.

[55] Ladde, G. S., Lakshmikantham, V., and Liu, P. T., Differential Inequalities and Stability and Boundedness of Stochastic Differential Equations. *J. Math. Anal. Appl.* **48** (1974), 341–352.

[56] Ladde, G. S., and Lakshmikantham, V., Stochastic Differential Inequalities of Itô Type. *Proc. Conf. Appl. Stochastic Processes, Athens, Georgia, May 14–19, 1978.* Academic Press, New York (to be published).

[57] Ladde, G. S., and Lakshmikantham, V., "Systems of Differential Inequalities and Stochastic Differential Equations V", (to appear).

[58] Ladde, G. S., and Leela, S., Instability and unboundedness of Itô type stochastic differential equations. *Rev. Roumaine Math. Pures Appl.* **XXII** (1977), 933–939.

[59] Lakshmikantham V., and Leela, S., "Differential and Integral Inequalities, Theory and Applications," Vol. I. Academic Press, New York, 1969.

[60] Leibowitz, M. A., Statistical behavior of linear systems with randomly varying parameters. *J. of Mathematical Phys.* **4** (1963), 852–858.

[61] Loéve, M., "Probability Theory," 3rd ed. Van Nostrand-Reinhold, Princeton, New Jersey, 1963.

[62] Lord, M. E., and Mitchell, A. R., A new approach to the method of nonlinear variation of parameters. *Appl. Math. Computation*, **4** (1978), 95–105.

[63] McLane, P. J., Asymptotic stability of linear autonomous systems with state-dependent noise. *IEEE Trans. Automatic Control* **AC-14** (1969), 754–755.

[64] McShane, E. J., "Stochastic Calculus and Stochastic Models." Academic Press, New York, 1974.

[65] Miyahara, Y., Ultimate boundedness of the systems governed by stochastic differential equations. *Nagoya Math. J.* **47** (1972), 111–144.

[66] Morozan, T., Stability of some linear stochastic systems. *J. Differential Equations* **3** (1967), 153–169.

[67] Morozan, T., Stability of linear systems with random parameters. *J. Differential Equations* **3** (1967), 170–178.

[68] Morozan, T., "Stabilitatea sistemelor cu Parametri Aleatori." Editura Academiei Republicii Socialiste România, Bucharest, 1969.

[69] Natanson, I. P., "Theory of Functions of Real Variables," Vol. I & II. Ungar, New York, 1964.

[70] Nevel'son, M. B., Stability in the large of trajectories of diffusion type Markov processes. *Differentcial'nye Uravnenija* **2** (1966), 1052–1060.

[71] Nevel'son, M. B., Some remarks concerning the stability of a linear stochastic system. *Prikl. Mat. Meh.* **30** (1966), 1124–1127.

[72] Nevel'son, M. B., and Khas'minskii, R. Z., Stability of a linear system with random disturbances of its parameters. *Prikl. Mat. Meh.* **30** (1966), 404–409.

[73] Nevel'son, M. B., and Khas'minskii, R. Z., Stability of Stochastic systems. *Problemy Peredachi Informatsii* **2** (1966), 76–91.

[74] Neveu, J., "Mathematical Foundations of the Calculus of Probability." Holden-Day, San Francisco, California, 1965.

[75] Prohorov, Yu. V., Convergence of Random Processes and Limit Theorems in Probability Theory. *Theor. Probability Appl.* **1** (2) (1956), 157–214.

[76] Ruymgaart, P. A., and Soong, T. T., A sample treatment of Langevin-type stochastic differential equations. *J. Math. Anal. Appl.* **34** (1971), 325–338.

[77] Samuels, J. C., On the stability of random systems and the stabilization of deterministic systems with random noise. *J. Acoust. Soc. Amer.* **32** (1960), 594–601.

[78] Samuels, J. C., Theory of stochastic linear systems with Gaussian Parameter Variations. *J. Acoust. Soc. Amer.* **33** (1961), 1782–1786.

[79] Shur, M. G., O lineinykh differentsial'nykh uravneniiakh so sluchaino vozmushchennymi parametremi (Linear differential equations with random perametric excitation). *Izv. Akad. Nauk SSSR Ser. Mat.* **29** (1964), 783–806.

[80] Skorokhod, A. V., Limit Theorems for Stochastic Processes. *Theor. Probability Appl.* **1**, (3) (1956), 261–290.

[81] Soong, T. T., "Random Differential Equations in Science and Engineering." Academic Press, New York, 1973.

[82] Srinivasan, S. K., and Vasudevan, R., "Introduction to Random Differential Equations and Their Applications." Amer. Elsevier, New York, 1971.

[83] Strand, J. L., Stochastic Ordinary Differential Equations. Doctoral Dissertation, University of California, Berkeley, California, 1967.

[84] Strand, J. L., Random Ordinary Differential Equations. *J. Differential Equations* **7** (1970), 538–553.

[85] Tsokos, C. P., and Padgett, W. J., "Random Integral Equations with Applications to Stochastic Systems." Springer-Verlag, Berlin and New York, 1971.

[86] Tsokos, C. P., and Padgett, W. J., "Random Integral Equations with Applications to Life Sciences and Engineering." Academic Press, New York, 1974.

[87] Wong, E., "Stochastic Processes in Information and Dynamical Systems." McGraw-Hill, New York, 1971.

[88] Wonham, W. M., Lyapunov criteria for weak stochastic stability. *J. Differential Equations* **2** (1966), 195–207.

[89] Wonham, W. M., Random differential equations in control theory. *In* "Probabilistic Methods in Applied Mathematics" (A. T. Bharucha-Reid, ed.), Vol. 2, pp. 132–212. Academic Press, New York, 1970.

[90] Yamada, T., On a comparison theorem for solutions of stochastic differential equations and its applications. *J. Math. Kyoto Univ.* **13** (1973), 497–512.

[91] Zakai, M., On the ultimate boundedness of moments associated with solutions of stochastic differential equations. *SIAM J. Control* **5** (1967), 588–593.

# INDEX

## A

Additivity, 4
Approximations,
  Carathéodory-type, 32
  Peano-type, 132
  successive, 43, 139
  stochastic, 119
Asymptotic stability, 70
  in probability, 70, 76, 78, 168, 171
  with probability one, 70, 82, 87, 90,
    177
  in $p$th mean, 70, 89, 110, 111, 174
Autocorrelation function, 181
Autocovariance function, 181

## B

Borel–Cantelli lemma, 7
Boundedness, 70
  in probability, 23
  sample, 23
Brownian motion, 14, 188

## C

Carathéodory-type condition, 71
Cartesian product, 2
Central limit theorem, 8
Chapman–Kolmogorov equation, 15
Chebyshev's inequality, 5
Closed graph theorem, 201
Comparison differential equations,
  deterministic, 19, 110, 166
  Itô-type or stochastic, 166
  with random parameters, 59

Comparison theorems,
  deterministic, 19, 108
  random, 41
  scalar version, 69, 160
  stochastic, 159
Compatibility condition, 12
Continuity,
  almost-sure, 23
  $L^p$-, 93
  in mean square, 93
  in probability, 22
  in $p$th mean, 93
  sample, 23
  strong, 93
Continuous dependence with respect to,
  initial data, 45, 108, 145
  parameters, 43, 105, 142
Convergence,
  almost surely or almost certainly, 6
  $D$-, 9
  in distribution, 6
  in mean square, 6
  in probability, 6
  in $p$th mean or moment, 6
  in quadratic mean, 6
  strong law of, 8
  weak law of, 8
Correlation-coefficient function, 181
Covariance, 5
Cross-correlation function, 181
Cross-covariance function, 181
Cylinder set, 2

## D

$D$-bounded, 9
$D$-Cauchy sequence, 9